SCIENCE AND TECHNOLOGY POLICY
Priorities of Governments

SCIENCE AND TECHNOLOGY POLICY
Priorities of Governments

C. A. TISDELL

Professor of Economics
University of Newcastle
New South Wales
Australia

London   New York
CHAPMAN AND HALL

First published 1981
by Chapman and Hall Ltd,
11 New Fetter Lane, London, EC4P 4EE

Published in the U.S.A. by
Chapman and Hall
in association with
Methuen, Inc.
733 Third Avenue, New York, NY 10017

© *1981 C. A. Tisdell*

Typeset by Inforum Ltd, Portsmouth

Printed in Great Britain
at the University Press, Cambridge

ISBN 0 412 23320 7

All rights reserved. No part of this book may be reprinted, or
reproduced or utilized in any form or by any electronic,
mechanical or other means, now known or hereafter invented,
including photocopying and recording, or in any information
storage and retreival system, without permission in writing
from the publisher.

British Library Cataloguing in Publication Data

Tisdell, C.A.
Science and technology policy
1. Science and state
2. Technology and state
I. Title
500   Q125   80-42128

ISBN 0-412-23320-7

Contents

Preface	ix
Abbreviations	xi

1 Basic Issues in Setting Priorities for Science and Technology Policy — **1**
1.1 Introduction — 1
1.2 The need for government involvement in science and technology and therefore priorities — 3
1.3 Goals and variables to be taken into account in science policy — 9
1.4 Goals for technology policy — 13
1.5 Centralization vs decentralization, comprehensiveness and the specification of priorities — 15
1.6 Forecasting and priorities — 18
1.7 Critical views about the role of science and technology in economic development — 19
1.8 Critical views of government support for science and technology — 23
   Notes and references — 26

2 Science Policy Options and Priorities — **31**
2.1 Introduction — 31
2.2 Education and the stock of knowledge — 33
2.3 Research and development – general issues — 38
2.4 Science and industrial policy — 40
2.5 Import of science vs its local supply — 46
2.6 Science and social policy — 52
2.7 Research for defence and big science — 55
2.8 Basic vs applied vs developmental science — 56
2.9 Concentration and dispersion of R & D effort — 59
2.10 Performers of R & D — 63
2.11 Service science — 66
2.12 Science and international affairs — 67
2.13 Some concluding comments — 71
   Notes and references — 71

3 Technology Policy: Options and Priorities 76
3.1 Introduction 76
3.2 Links between science and technology and technology sequences 77
3.3 Inventions 80
3.4 Innovations 83
3.5 Diffusion of new technology 86
3.6 Replacement of equipment 90
3.7 Domestic technology transfer 91
3.8 International transfers of technology 93
3.9 Environmental overspills and technology 96
3.10 Employment problems and other social aspects of technology 97
3.11 Observations 100
   Notes and references 101

4 Science and Technology Policy in Large OECD Economies 106
4.1 Introduction and background data 106
   Federal Republic of Germany
4.2 Articulation and administration of priorities in West Germany 110
4.3 Selected features of German Science and Technology priorities 113
   Japan
4.4 Articulation and administration of priorities in Japan 118
4.5 Selected features of Japanese science and technology priorities 123
   United Kingdom
4.6 Articulation and administration of priorities in the United Kingdom 126
4.7 Selected features of United Kingdom science and technology priorities 131
   United States of America
4.8 Articulation and administration of priorities in the United States 134
4.9 Selected features of American science and technology priorities 140
4.10 Some observations 143
   Notes and references 145

5 Science and Technology Policies of Small OECD Economies 149
5.1 Background 145

Belgium
5.2 Articulation and administration of priorities in Belgium 152
5.3 Selected features of Belgian science and technology priorities 154
Canada
5.4 Articulation and administration of priorities in Canada 157
5.5 Selected features of Canadian science and technology priorities 161
The Netherlands
5.6 Articulation and administration of priorities in the Netherlands 164
5.7 Selected features of Dutch science and technology priorities 168
Sweden
5.8 Articulation and administration of priorities in Sweden 170
5.9 Selected features of Swedish science and technology priorities 175
Switzerland
5.10 Articulation and administration of priorities in Switzerland 177
5.11 Selected features of Swiss science and technology priorities 181
5.12 Some observations 183
Notes and references 185

6 Retrospect and Prospect 188
6.1 The increased emphasis on priority assessment in science and technology policy 188
6.2 Macro approaches to taking account of science and technology priorities 190
6.3 Efficiency and science and technology priorities within sectors 191
6.4 Changing science and technology priorities 192
6.5 Why the trend towards co-ordination and explicit priorities in science and technology policy? Fundamental reasons 193
6.6 Problems inherent in the basic trend 194
6.7 Problems inherent in observed government priorities 198
6.8 In conclusion 202
Notes and references 204

Index 206

Preface

I was asked recently to prepare an independent background report on the subject of priority assessment in science and technology policy for the Australian Science and Technology Council. The Council (while not necessarily endorsing this book) suggested that a wider audience could be interested in the type of material contained in my report and kindly gave me permission to publish the material in my own right. The present book contains this and other material, some of which was presented at a seminar on *National Science Policy: Implications for Government Departments* arranged by the Department of Science and the Environment. Additional ideas were developed in response to comments on the manuscript by referees, as a result of discussions with Professor John Metcalfe and Dr Peter Stubbs of Manchester University, a conversation with Dr Keith Hartley of the University of York and in the wake of a communication from Dr Ken Tucker, Assistant Director, Bureau of Industry Economics, Australia.

Science and technology policy affects and concerns everyone of us if for no other reason than we cannot escape in this interdependent world from the economic, social and environmental overspills generated by science and technology. We must face the problems and promises inherent in new and existing science and technology whether we like it or not. Not surprisingly this book finds that all industrialized countries seem to be facing similar economic and social problems. Scientific and technological change on the one hand may add to these problems and on the other hand promises redemption from them. This poses the perplexing problem for communities of deciding on the best type of science and technology policy for governments to follow. What options should be considered and what priorities should be adopted by governments in formulating their science and technology policies? What priorities have been adopted in practice?

The first three chapters of this monograph provide an analytical review of priority assessment in science and technology policy. The next two chapters report on the science and technology policies and priorities of nine selected OECD countries. Chapter 6 comments on broad trends in science and technology assessment in the light of the individual country reviews. In reviewing priority assessment in science and technology policy in the selected countries, it has of course only been possible to give a general picture. Nevertheless, this type of picture seems appropriate for a broad comparative analysis of science policy.

A reader wishing to avoid analytical detail (this detail is essential for appreciating the full ramifications of priority assessment in science and technology policy), might wish to be selective in browsing through Chapters 2 and 3; they should concentrate principally on the introductory sections. The reading of Chapters 2 and 3 should not be regarded as a prerequisite for following the content of Chapters 4, 5 and 6 but these chapters provide greater depth to the study.

This is an independent study and the views expressed in it (and errors if any) are solely the responsibility of the author. But I would like to place on record not only my thanks to Mr Bruce Middleton, Assistant Secretary, ASTEC, for his assistance in providing me with access to source material and for his encouragement with the original project, but the general helpfulness of the ASTEC secretariat. I appreciate also the assistance with source material provided by diplomatic representatives of overseas missions in Australia. Without such assistance the study would be of diminished value.

Mrs Lorraine King and Miss Elizabeth Williams deserve particular thanks since they typed the whole manuscript and their efficiency and pleasant nature made my task easier.

My family was very understanding of my attempts to keep on schedule with this manuscript and this was of considerable assistance.

Clem Tisdell

Charlestown, 2290
Australia
December, 1980

Abbreviations

ABRC	(UK) Advisory Board of the Research Councils
AIST	(Japanese) Agency for Industrial Science and Technology
ACARD	(UK) Advisory Council on Applied Research and Development
ARC	(UK) Agricultural Research Council
BMFT	(German) Ministry of Research and Technology
CERN	European Organisation for Nuclear Research
CST	(Japanese) Council for Science and Technology
DFG	German Research Association
DOD	(US) Department of Defence
DOE	(US) Department of Energy
EEC	European Economic Community
ERP	European Recovery Programme
ESA	European Space Agency
FFRDC	(US) Federally Funded Research And Development Centres
FOA	(Swedish) National Defence Research Institute
FRN	(Swedish) Council for Planning and Co-ordination of Research
GERD	Gross Expenditure on Research and Development
GDP	Gross Domestic Product
GNP	Gross National Product
HEW	(US) Department of Health, Education and Welfare
IRIC	(Canadian) Industrial Research and Innovation Centres
JRDC	Japan Research and Development Corporation
LDC	Less Developed Country
MEW	Measurable Economic Welfare
MITI	(Japanese) Ministry of International Trade and Industry
MOSST	(Canadian) Ministry of State for Science and Technology
MPG	Max-Planck Society for the Promotion of Science
MRC	(UK) Medical Research Council
NASA	(US) National Aeronautics and Space Administration
NATO	North Atlantic Treaty Organisation

NERC	(UK) National Environment Research Council
NNP	Net National Product
NRC	(Canadian) National Research Council
NRDP	(Japanese) National Research and Development Programmes
NRDC	National Research and Development Corporation
NSF	(US) National Science Foundation
OECD	Organisation for European Co-operation and Development
OSTP	(US) Office of Science and Technology Policy
OTA	(US) Office of Technology Assessment
PPB	Planned Programme Budgeting
R & D	Research and Development
S & T	Science and Technology
SAREC	Swedish Agency for Research Co-operation with Developing Counties
SIDA	Swedish International Development Authority
SRC	(UK) Science and Research Council
SSRC	(UK) Social Science Research Council
SST	Supersonic Transport
STA	(Japanese) Science and Technology Agency
STEP	(Canadian) Science and Technology Employment Programme
STU	National Swedish Board for Technical Development
TNO	(Netherlands) Organisation for Applied Scientific Research
USDA	US Department of Agriculture
ZWO	(Netherlands) Organisation for Advancement of Pure Research

CHAPTER ONE

Basic Issues in Setting Priorities for Science and Technology Policy

1.1 Introduction

Government management or control of science and technology has increased and governments have become more concerned about their formulation of science and technology policy. The reasons for this are complex and no doubt subject to dispute. Nevertheless, a number of observations appear in order. While the community at large still looks to advances in science and technology as a means of improving the lot of mankind, it questions the social benefits of unbridled scientific and technological change. Indeed many members of the community have become fearful of the possible unwanted effects of technological change, for instance nuclear risks, unemployment, global pollution. To gain the maximum benefits from scientific and technological change and to avoid unwanted environmental and social consequences, there has been a growing community demand for science and technological effort to be more closely supervised through the government to meet social goals. The belief has gained ground that the direction of scientific effort should not be left to scientists, technocrats or even business managers acting alone but that government reflecting community-wide interests should play a greater role in directing technological change. Specific concerns such as defence, then environmental damage and the depletion of non-renewable resources and, more recently, increased international economic competition accompanied by economic recession, have brought demands for improvements in the science and technology policies of governments.

Apart from these pressures, however, it is necessary for governments in modern economies to give greater consideration to goals or priorities and efficiency in their science and technology policies because they are responsible for a high proportion of the funding of science and technology expenditure in capitalist and quasi-capitalist countries and are active performers in the educational and research development fields.[1] By their policies of various kinds, whether well designed or randomly formulated, governments also influence the performance of individuals and companies in adding to science and technology and employing it irrespective of whether governments partially fund this activity.

However, for the government to manage science and technology in the social interest is not an easy task. Indeed, some may despair about the possibilities of such management because the social interest is not always clear cut and the government is a part of the political scene (not above it) and therefore subject to whatever imperfections exist in the system. One can only agree with King that:

> The management of technology in the broad social interest is a very complicated process which must take into account social as well as economic costs and benefits and foresee the long-term effects of its achievements over the broad spectrum of human activity. For its optimum use, goals must be more clearly formulated than at present and, until this has been done, technology management can only hope to avoid major disutility, while attempting to achieve its direct objectives.[2]

The purpose of this monograph is twofold. It isolates economic and, to a lesser extent, social factors that ought to be taken into account in formulating and planning science policy from society's point of view. It addresses such questions as the following: What economic and social grounds exist for government intervention in scientific and technological change? What factors constrain or limit the possibilities for government management of science and technological change? How best can policy formulations be organized and administered in the light of these constraints? What economic and social goals, priorities and trade-offs need to be assessed by government policy-makers in formulating science and technology policy? These matters are dealt with in the first part of the monograph which brings together in an analytical fashion a variety of views that have previously been scattered and unrelated. Some of these views (for example on international economic competition and science) heavily influence policy-makers as we shall see even though these views have not yet become established in mainstream economic theories.

The second purpose and part of the monograph is to review the science and technology policies that have in fact been adopted by selected OECD countries. Five small economies (Belgium, Canada, Netherlands, Sweden and Switzerland) and three large economies (United Kingdom, United States and West Germany) are selected for specific consideration. It is intended to summarize the way in which science and technology policy is formulated, planned and administered in these countries and to pay particular attention to the priorities that have emerged. This enables common trends and differences between the countries to be isolated and discussed and the following questions to be considered: Why have the common trends and differences emerged? To what extent have the countries concerned adopted or acted upon the principles outlined in the first part of the monograph? Are we justified in placing so much hope in government science and technology policy as a means to improve the lot of mankind?

1.1.1 *Distinction between science and technology*

The terms 'science' and 'technology' are widely used but there is some dispute about their precise meaning. In practice, science and technology overlap and there is little point in attempting rigid definition. The *Concise Oxford Dictionary* defines 'science' as systematic and formulated knowledge and 'technology' as the science of the industrial arts. However, 'technology' has also a wider connotation and refers to the collection of production possibilities, techniques, methods and processes by which resources are actually transformed by man to meet human wants.[3] I shall take this broad view implying as it does considerable overlap between science and technology.

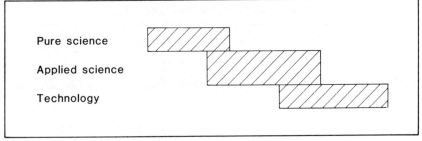

Fig. 1.1 Science and technology overlap.

Incidentally, Feibleman's views[4] of the relationship between pure science, applied science and engineering are pertinent. In his view there can be no applied science without pure science but there can be pure science without applied science and without follow-through to the technological stage, as for instance happened at one time in ancient Greece. While in modern society applied science is the source of most technology, it is in Feibleman's view possible to develop some technology without science by trial and error. Possibly early man relied to a great extent on trial and error as a basis of his technological development but modern man is unable to rely to any great extent on this method if he is to continue the momentum of his development of technology.

1.2 The need for government involvement in science and technology and therefore priorities

Economists have identified a number of reasons in favour of government interference in scientific effort, technological development and use of technology. The arguments in certain cases support government funding or assistance for scientific and technological activities and in some other cases government guidance and/or limitations or restrictions on private decisions about scientific effort and the use and development of technology.

1.2.1 *Economic reasons for government intervention in scientific effort*

There are a number of reasons, summarized in Table 1.1, why the majority of economists believe that a free-enterprise market system in the absence of government intervention is likely to support insufficient scientific effort and not allocate this effort in an efficient pattern. Markets fail to work as sufficiently adequate mechanisms for allocating resources to scientific effort because decisions made by firms and individuals are based upon their private profits and gains and these frequently differ from social gains.

Table 1.1 Summary of some economic reasons for government interference in private scientific and technological effort.

---

1. Individuals or individual companies unable to appropriate adequate share of total gains from such efforts.

2. Risks and uncertainties associated with such efforts not adequately taken into account by private agents.

3. Social failures in the transmission of scientific and technological information. Associated backward and relatively ignorant groups.

4. Imperfections in capital markets, that is in the provision of funds for scientific effort and technological change.

5. Avoidance of wasteful duplication of scientific services.

6. Considerations of national security.

7. External industry-wide economies of development coupled with the failure of markets to co-ordinate and direct some large-scale desirable initiatives. Gives rise to a selective field or selective industry policy approach.

A company or an individual may *not* be *able to appropriate an adequate share of the total gains* to society from a discovery to compensate him for the cost of making that discovery even though the total gains to society significantly exceed the cost of making the discovery.[5] Thus, if an individual develops a commercially valuable new species of pasture he cannot, once it has been released, prevent others from using it and they can use it without compensating him in any way for his effort. His gains, if any, may be limited to his initial sales. Where a basic new principle is discovered which is subsequently shown to be of value to industry the same problem occurs.

The patent system is intended partially to remedy this situation, but the system cannot be applied to all discoveries (it cannot be applied to discoveries in basic science, for instance, because use of 'the property' by others cannot be easily discerned). The system raises a number of difficulties of its own because it provides a temporary monopoly and in some cases monopoly power to an inventor, and it does not always ensure adequate appropriation for the inventor or discoverer. In the last respect, for instance, the property rights of inventors are curtailed in time and they have no rights in inventions which may stem from the original invention. The issuing of, conditions attaching to and use of patents and legal property rights in discoveries and inventions have been and are thorny issues in most countries. While patents permit some appropriation of benefits by some discoverers as a reward for their scientific effort, they do not ensure adequate compensation for all scientific effort and can have undesirable social side-effects.[6]

Risks and uncertainties provide grounds for government intervention in scientific effort and technological development. Two different bases for interference seem to exist: (i) to counteract risk-aversion by individuals and companies where this is appropriate and (ii) to ensure that proper account is taken of collective risks to society.

While results from individual scientific or technological efforts are risky or uncertain, overall results from the scientific and technological efforts of all are less risky or uncertain provided that individual results are not perfectly correlated. The variance or dispersion of *pooled* results is smaller as a rule than that for individual results. It has been argued by Arrow that society should make its decision to commit funds to scientific effort upon the basis of the pooled situation.[7] This as a rule indicates that it is desirable to allocate more resources to scientific effort than would be done by risk-averse individuals operating in isolation on the basis of their own results, that is on a non-pooled basis.

Not all risks and uncertainties of scientific and technological change are borne by the initiators of these nor are the initiators always required or able to compensate other parties at risk or actually damaged. For instance, suppose that a private company in building and operating a nuclear installation places many people (members of the public) at risk of serious injury or death. The company is unlikely to pay compensation to these individuals for placing them at this risk. Further, should a mishap occur and damages be payable, an injured member of the public may believe that no monetary payment can adequately compensate him for serious injury let alone for death. This aside, however, the liability of a company must always be limited either by its available assets or the limited liability of its shareholders. Therefore where damages are very great the company may be shielded to some extent from the full damages occasioned. The matter of compensation is further complicated when the risks and uncertainties associated with scientific and technological change are remote and difficult to trace as can occur with variations in the frequency and type of genetic mutations resulting from technological change. Collective risks can easily differ from the risks borne by initiators of scientific and technological change. Since initiators can be expected only to take account of their own risks, a government needs to interfere to ensure that adequate heed is taken of collective risks when the two types of risks differ substantially.[8]

Of course it is also possible that those in the vanguard of scientific and technological change may not adequately assess their own risks. The action which a government ought to take in this case is much more controversial but in some cases government acts to impose safety restrictions, for instance limitations on radiation doses to be received by researchers.

Individuals and companies can be too myopic (short-sighted) on occasions to adopt or develop new technology even when they can expect substantial long-term gains from developing or adopting new technology. They are unprepared to make short-term sacrifices for long-term gains, that is invest in development and bear the disruption and 'teething troubles' incurred in installing a new technology. Some economists believe that there is a case for the government to intervene to counteract serious myopia of this kind.

Some *groups in society* tend to be *technologically backward*. There are many possible reasons for this. In some cases cultural background and attitudes to change may play a role and in other cases, the group may have had an unfortunate experience in adopting new technology which in the event proved to be inferior to earlier technology for them. However, for individuals comprising some groups, it is as a rule uneconomic *for them* (as individuals) to be in the vanguard of scientific development and new technology. Other things equal, the smaller a firm is, the less it pays it to spend on collecting and processing new knowledge. For instance, suppose that a company (firm) consisting of one small plant can improve its annual profit by $100 per annum if it searches for, discovers and adopts a suitable technique. A company operating ten such plants (in a linear world) would, if it discovered and adopted the technique, increase its annual profit by $1000. It pays the larger firm, as a rule, to spend more on scientific effort and on the collection of knowledge about new technology. This helps to explain why small firms are frequently technologically backward.

On economic efficiency grounds, there is a case for the government assisting with the development of scientific knowledge of particular relevance to *small firms* and for the government assisting in the dissemination of technological knowledge to this group. Extension services to farmers provide one example of this activity at present. Although no individual may find it worthwhile to collect information and disseminate it, it may be socially worthwhile for the government to do so. For instance suppose that there are 1000 firms without a piece of information which, if they had it and acted on it, would add $10 each to their profit and give $10 gain to others in society for each firm using it. The costs of collecting the information, say $50, makes it uneconomic for any firm to do this, if we assume that the information cannot be effectively marketed to other firms. The information is not used. However, suppose that a government agency collects the information at a cost of $50 and disseminates it to all the firms at a total cost of $2000. Society gains by the equivalent of $17 950, that is by the $10 000 increase in profit to firms plus the $10 000 gain to other members of society less the cost of $2050 of collecting and disseminating the knowledge.

Capital markets, that is markets supplying funds for investment and other purposes, are claimed to be *imperfect* sometimes. Since most scientific and technological change requires investment these activities can be adversely affected. At least two types of problems can be identified: (i) small firms may have restricted access to funds and this can retard their technological performance; (ii) the capital required for some research and development projects is so large, the length of the investment before returns are obtained and the uncertainty of returns is so great, that funds cannot be easily raised in the free market and continuity of their supply cannot be assured. Imperfect capital markets may form a basis for government intervention in scientific and technological effort. The decision of governments to invest in civilian big science has been attributed by Pavitt and others to the inability of free markets to finance these investments.[9]

The government may interfere in the provision of scientific services to avoid *wasteful duplication* of these. Duplication of standards, meteorological services and other services might be expected to add to their cost without improving the service provided.

In cases of *national security,* cases in which there are large spill-over ramifications for the future of society, governments may need to intervene in scientific and technological development. While in the case of military ability this has long been recognized, diplomatic economic and social defence could also require such interference. This has become increasingly recognized as relative supplies of oil have dwindled and the economies and societies of nations dependent on oil and other energy imports have become increasingly subject to policies in oil-exporting countries.

Markets may fail to steer industry and associated research and development into areas or industrial fields of greatest national gain or advantage because decisions about resource use are made upon a basis that is too individualistic and isolated.[10] While it may not be profitable for an individual firm to branch out into a new field the position may be quite different if many firms can be convinced to enter that field almost simultaneously. They may then obtain external economies and economies of agglomeration. External economies (dependent upon the development of the whole industrial field)[11] may be achieved because specialists spring up to deal with different parts of the activities of the industry, industry overheads such as the promotion of the industry abroad can be spread over a greater output and advantage of any research results obtained can be taken advantage of by a larger group. The problem and the mechanism is somewhat similar to that envisaged in the development of new regional towns and cities.[12] The implication of this view is that the government may need to co-ordinate and encourage *the development of selected industries* or selected industrial fields and back its industrial strategy by giving support to science and industrial research and development and technological advance appropriate to the selected industries or fields. This approach further ties in with the idea that the most significant profit to be made from international trade is from being a technological leader or near-leader in an appropriate field. Japanese industrial policy has been to a large extent based on this point of view.[13]

Policies of this kind are often accompanied by selective protectionism. Protection from foreign imports may be afforded to infant industries or infant industrial fields until effective experience and learning have built-up and the scale of the *whole* industry or field has reached a stage where the home industry is more than competitive. The government may also interfere in technological exports to ensure maximum gains for the home country and that imports of technology are obtained on the best possible terms from a national point of view.

While selective industrial, science and technology policy may bring significant national gains, it involves social risks. A selected industry may prove to be uneconomic in the long term and negative or poor returns may be obtained on the large national investment involved in fostering it. While this suggests caution in considering this approach, it does not provide a case for rejecting it without serious consideration. The argument basically revolves around whether the selection of industries for development should be left to market forces or whether a case exists for the government to interfere and co-ordinate and encourage the development of specific industries. Will free markets ensure the best industrial specialization for a country? If not, how can the best specialization be identified and can governments be expected, given the political constraints upon them, to guide the economy towards its best industrial specialization?

Pavitt in discussing governmental support for research and development in France provides a number of illustrations of economic justification for governmental support and gives a useful summary of the main points in favour of such support, despite his 'unsympathetic' view of French science policy.[14] However, his paper is useful in so far as it emphasizes the point that in most cases the alternative to imperfect markets tends to be imperfect rather than perfect government intervention. Policy involves a choice between imperfect alternatives.

1.3 Goals and variables to be taken into account in science policy

1.3.1 *Complexity of goals*

Not all of scientific effort is directed towards improvements in the tangible welfare of human beings. Much scientific effort is aimed at satisfying human curiosity, the need to know, explain and relate to things around us. Man does not live by bread alone. There is no doubt that most men value some knowledge for its own sake and are prepared to make material sacrifices to obtain such knowledge.

This is not to deny that the goal of possibly most scientific effort is to improve the standard of living of human beings, but this is not the only legitimate goal of science.

However, it is difficult to measure human welfare or the standard of living precisely. For instance, if higher material wealth is accompanied by greater tension and stress in the population, is welfare increased? If the welfare of one group is increased by a scientific advance and that of another reduced, is the change desirable on balance and how is its desirability to be decided?

1.3.2 *Some alternative 'official' approaches to classifying research and science objectives*

The OECD in following the Planned Programme Budgeting (PPB) method distinguishes *national goals* which are 'statements of highly desirable conditions towards which society should be directed' and *objectives* which are 'the stated purpose of an organisation or an individual capable of planning and taking action to gain intended ends'.[15]

In its view 'the formulation of national goals is a political process involving all elements in our society, especially the makers of public opinion and the political leadership' and is too broadly defined to suffice for R & D analysis. Because of this, functional analysis of government R & D is usually undertaken by *objective* even though there is not common agreement about how objectives should be defined.

The OECD has identified three main approaches to defining R & D objectives: (i) purpose; (ii) area of relevance and (iii) intended result.[16]

The purpose approach to government funding of R & D is linked to apparent purposes as expressed in national budgets but 'it implies that governments know in advance the functional allocation of R & D appropriations in their budgets' otherwise these have to be 'assessed'. Where a project has multiple objectives it is usual in purpose-type analysis to assign it in accordance with the dominant objective. No allowance is made for 'spin-off' or secondary objectives.

The area of relevance approach looks more carefully at the purposes for which R & D is performed and makes allowance for secondary objectives and spin-off. The detail required for this approach can only be obtained by surveying regularly performers of R & D and this can be costly. Only Belgium and Canada have adopted this approach. As with the other approaches identified by the OECD, the relative importance of different R & D objectives is measured by the *amount of funds* allocated to them.

The 'results' approach considers the division of R & D funds between the end-products or missions to which they directly or indirectly contribute and takes this composition as an indicator of the weight placed on different objectives. The end-product or mission may be a new or improved product or a new or improved technical process. The OECD points out that 'its use requires the formulation of an extremely detailed classification and also widescale surveying. It is probably best fitted for case studies within a given field of relevance'.

1.3.3 *The classification of research and science objectives used by OECD and EEC*

The OECD groups its R & D objectives into:
  national security and big science;
  economic development;
  community services;
  advancement of science;
  other activities.
The EEC groups its objectives for government funding of science into:
  defence and advanced technologies;
  agricultural and industrial purposes;
  social purpose;
  general advancement of knowledge;
  other.
Each of the OECD and EEC categories is further subdivided in terms of objectives as set out in Tables 1.2 and 1.3. The general approach of these bodies to considering objectives is a functional one.

Table 1.2 OECD classification of objectives for research and development policy.

Group	Objective	Sub-class
National security and big science	Defence Civil nuclear Civil space	 Civil nuclear only Civil space only
Economic development	Agriculture, forestry and fishing	Agriculture, hunting, fishing and foerstry
	Mining and manufacturing	Mining and manufacturing industries including computers
	Economic services	Construction and public works Communications Transports Gas, electricity and water
Community services	Health	
	Pollution	
	Public welfare	Education Social services Urbanism Culture and recreation
	Other community services	Disaster prevention Law and order Meteorology Planning and statistics
Advancement of Science	Advancement of Research via general university funds	
Other activities	Developing countries	
	Miscellaneous	

Source Based on p.85 OECD (1975), *Changing Priorities for Government R & D*, Paris.

Table 1.3 EEC classification of objectives for government funding of science.

Group	Objective	Sub-class
Defence	Defence	
Advanced technologies	Nuclear	Civil only
	Space	Civil only
	Computer science and automation	
Credits for agricultural and industrial purpose	Agricultural productivity	Agriculture, hunting, fishing and forestry
	Industrial productivity	Manufacturing industries
Credits for social purpose	Human environment	Construction Civil engineering Transport Telecommunications
	Earth and atmosphere	Soil + sub-strata (mining industries) Seas and oceans Atmospherics incl. meteorology
	Health	Medical research Hygiene and nutrition Pollution
	Social services and humanities	Education training Readaption Business Management
General advancement of knowledge	Outside higher education	
	Higher education	
Not broken down elsewhere		

Source Based on p.85, OECD (1975), *Changing Priorities for Government R & D*, Paris.

Limitations
While the availability of statistics and the costs of collecting and processing statistics as well as time-lags in collection limit the most practical approach

to classifying research and science objectives, it is appropriate to make a number of observations about the OECD approach. In practice, objectives tend to overlap between categories. In big science, for instance, civil nuclear overlaps with economic development. Community services, such as education and health, may also contribute to economic development.

Even at their most disaggregated level (as set out in the third column of Tables 1.2 and 1.3) the OECD and EEC categories are too broad for planning science expenditure. For instance, the overall level of support for research and development expenditure in mining and manufacturing is insufficiently discriminating as an objective. It is necessary to consider what kinds of industries or fields are to be supported in mining and in manufacturing industry and why? The OECD classification at best gives a broad framework in which changes of apparent emphasis on R & D objectives can be gauged on the basis of variations in expenditure between the selected R & D categories.

1.4 Goals for technology policy

The development of new technology, mostly as a result of scientific effort, and its use has been a powerful factor in raising human welfare. New technology has contributed greatly to economic growth by raising Gross Domestic Production (GDP) and has enabled a greater population to live at a higher standard of living than otherwise possible.[17] In the broad, it has also contributed to an improvement in the health of people, increased the amount of leisure available to them, made improved working conditions possible and in most cases a reality, and has provided means to reduce pollution and improve environmental conditions. Given the current expressed views, especially as far as the desirability of economic growth is concerned, one could easily be misled into believing otherwise.

Although some technology may be developed to satisfy human curiosity, most is aimed at tangibly increasing human welfare. Just how does one measure human welfare? The recent debate about growth in GDP and increased welfare brings out the difficulty of obtaining a suitable index of human welfare and identifying the factors contributing to it.

Improved technology has and is a significant element making for the growth of GDP and in the 1950s and 1960s this was regarded as a most significant benefit of technological and scientific progress. The 1970s have seen much questioning of the desirability of economic growth and of new technology to support such growth[18] and this concern is bound to continue even though emphasis on it may alter.

GDP is not an adequate indicator of social welfare because:

(a) Many of the components of GDP are social costs rather than benefits. For instance, GDP increases if pollution rises and people are employed to clean up pollution or if travelling time to work increases because of greater congestion, thereby raising fuel consumption.

(b) GDP does not make an allowance for all desired and consumed goods. It allows only for traded goods. No allowance (or only allowance at cost) for instance is made for public recreation areas, for the amount of leisure enjoyed or improved conditions in the work place just as no allowance is made for deteriorating environmental quality.

(c) The allowance made for some goods is less than their relative value. For instance, to include health services at cost probably seriously understates their value.

Because of the limitations of GDP as an indicator, economists have attempted to develop other indicators of welfare. For instance, Nordhaus and Tobin have developed a concept which they call 'measurable economic welfare' (MEW) which involves additions to and subtractions from Net National Product (NNP).[19] They make an addition for leisure and non-market activities and subtract an allowance for 'regrettable necessities' and 'disamenities'.

In turn MEW can be subjected to considerable criticism as a measure of economic welfare.[20] The subtraction for 'regrettable necessities' in particular is debatable. Regrettable necessities are items which we need but regret having to have. Nordhaus and Tobin regard defence expenditure as one such necessity but they do not dismiss the possibility that new technology may lead to the engineering of wants or ends and raise expectations which cannot be met and so reduce happiness.

It is clear that new technology can add to economic growth, reduce pollution, improve the quality of life and help meet our wants. The anti-growth school's arguments are not able to detract from the importance of scientific and technological progress as a means to attaining human aims.

In relation to the growth question itself Beckermann has made the following point:

> First, much of the alleged 'costs' of economic growth, such as excessive pollution of the environment or inadequate attention to the public sector, are really nothing to do with the costs of growth. The growth issue is a problem of the allocation of resources over time, not of the allocation of resources at any moment of time, except in so far as the optimum growth rate requires the optimum allocation of resources to investment at any moment of time. Pollution and the like are instances of misallocation of resources at any moment of time, and there is no reason to believe that a slower rate of growth would help eliminate this. Indeed, it is likely to have the reverse effect, for it is easier to improve resource allocation in a growing economy than in a stagnant one. A change in resource allocation invariably means a change in income distribution. It is hence easier to carry out the former if the inevitable change in income distribution can also be carried out in the context of a growing

economy, since this enables the relative shares of some members of the society to be reduced without their absolute levels of income being reduced.

Secondly, the facts relating to many of the most important ingredients in the quality of life in advanced countries, such as the environment, health conditions, or leisure and working conditions, show that they have all improved markedly during the course of the last few decades, not to mention during the course of the last century.

Thirdly, it is argued that one of the popular criticisms of economic growth – namely that the increased consumption is merely the response to rising 'needs', many of them 'artificially' induced – is based on an oversimplification of the philosophical issues involved.[21]

No matter whether growth of GDP is a prime objective or other aspects such as environmental conditions, working conditions, health or leisure are important, appropriate technology can help to achieve these objectives more effectively. In this respect, priorities need to be established:

(a) for the encouragement of innovation;
(b) for the diffusion of new technology;
(c) for the replacement of technology;
(d) to take account of the spill-overs or side-effects of alternative technologies;
(e) on whether to import new technology rather than produce it at home;
(f) on whether to be up-to-date in a few technologies even if this is at the expense of others.

These aspects are taken up in Chapter 3 of this monograph.

1.5 Centralization *vs* decentralization, comprehensiveness and the specification of priorities

Consideration needs to be given (a) to the social method by which science and technological priorities are set, (b) to centralization and decentralization in decision-making about scientific and technological effort, and (c) to the extent to which it is desirable to plan priorities and make these comprehensive. Furthermore, one must consider the status of stated goals. Are they merely exhortative? If not, what mechanisms are used to ensure that they are followed? Indeed, these matters raise the whole question of the extent to which national scientific and technological *planning* is desirable.

On the desirability of national planning of scientific and technological effort and the degree of desirable planning, there are divergences of opinion and practice. In line with their policy of indicative economic planning, French scientific and technological policy is relatively centralized whereas in contrast the approach of the United States is much more pluralistic.[22]

It is difficult to classify science and technology policies in terms of the degree to which they involve central decision-making and control because there are many characteristics of policy-making which can vary in degree of centralization. However, it is worthwhile from a theoretical point of view distinguishing between the following two types of systems:

(a) Those attempting a rational and comprehensive approach with overall goals consistently specified and centrally distilled; decision-making is centrally co-ordinated (if not centrally directed) in accordance with a theory or model of the operations of society and its interrelationships.

(b) Those systems relying upon the interactions of groups with different goals and with limited perspectives; in these systems there is no overall co-ordination nor need there be agreed common goals.

Lindblom is one of the chief opponents of the rational-comprehensive approach and favours the interaction non-centrally co-ordinated approach.[23] Arguments against the rational comprehensive approach are:

(a) It is impractical because the amount of information to be collected and processed to apply it is immense, costly to collect and process and beyond the power of any individual to comprehend in depth given the limited capacity of men.

(b) Common collective goals do not always exist and goals formulated from above or at a centralized place may get out of step with those at the grass root level.

(c) The approach can result in 'large leaps being made in the dark' on the basis of unsubstantiated theory and is therefore dangerous.

On the other hand the advantages of the interaction root-type system are claimed to be:

(a) It allows many to participate in decision-making and takes account of the diversity of goals and experience in society.

(b) The interaction of the various groups with their own experiences to bring to bear means that greater use is made of available information and experience and it is possible that collective goals may be more likely to be achieved than in the centralized case.

(c) There are less likely to be 'leaps into the dark' because in this system only small or marginal policy adjustments are likely to be considered.

Whether or not the last mentioned 'stability' characteristic is desirable is likely to depend on one's own point of view. Some may believe that a system of this kind excessively hampers social innovation whereas others may be glad to have a dampener.

Lindblom's point of view on policy-making, if one is convinced by it, strikes at the very foundation of national priority assessment for science and technology policy intended to improve choices between alternatives, for:

> Lindblom's point of departure is a denial of the general validity of two

assumptions implicit in most of the literature on policy-making. The first is that public policy problems can best be solved by attempting to understand them; the second is that there exists sufficient agreement to provide adequate criteria for choosing among possible alternative policies. Although the first is widely accepted – in many circles almost treated as a self-evident truth – it is often false. The second is more often questioned in contemporary social science; yet many of the most common prescriptions for rational problem solving follow only if it is true.

Conventional descriptions of rational decision making identify the following aspects: (a) clarification of objectives or values, (b) survey of alternative means of reaching objectives, (c) identification of consequences, including side-effects or by-products, of each alternative means, and (d) evaluation of each set of consequences in light of the objectives. However, Lindblom notes, for a number of reasons such a *synoptic* or comprehensive attempt at problem solving is not possible to the degree that clarification of objectives founders on social conflict, that required information is either not available or available only at prohibitive cost, or that the problem is simply too complex for man's finite intellectual capacities.[24]

Degrees of comprehensiveness can vary, however, and it is possible to have mixed systems. Indeed, most policy-making systems in practice tend to be mixed with elements of centralization and comprehensiveness as well as decentralized components and areas in which comprehensiveness is absent. This is why it is difficult to classify policy-making systems. Suitable mixed or intermediate systems, involving partial specification of goals and partial centralization, may combine the best features of the two extreme types discussed above.

Figure 1.2 schematically illustrates three alternative degrees of centralization in priority setting for a government body as far as its spending on science is concerned. The size of each square represents the size of the body's budget. In case (a) it is free to spend its budget however it pleases (provided that it carries out its general charter). In case (b) a central controlling authority earmarks an amount of funds to be spent on scientific effort by the recipient government body but does not allocate this spending by areas. In case (c) an amount of funds is earmarked for scientific effort and divisions between scientific fields or areas are specified or imposed by the central authority on the recipient body. The greater the number of these restrictions, the more centralized is the system and the more likely that some of the difficulties mentioned by Lindblom will become a reality.

In systems which are not highly centralized advisory science bodies can play a useful co-ordinating role by providing information and ideas about the scientific policies pursued by different bodies. Large deviations in policy by bodies from 'the desirable' for instance can be brought to the attention of the central authority and action can be taken to modify such deviations. Areas of neglect, science and technology areas not served by any body, can

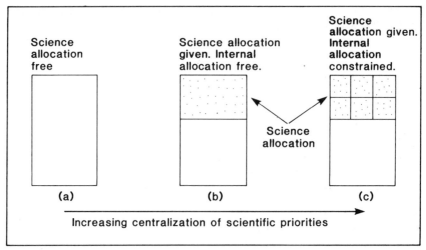

Fig. 1.2 Budget allocation to a government body from a central authority showing different degrees of specification of priorities for scientific effort.

also be brought to the attention of the central authority by the advisory body so that the central authority can take action to remedy the situation. This system allows for management by exception.

1.6 Forecasting and priorities

In setting priorities it is as a rule necessary to engage in forecasting about what is possible or is likely to occur. At least this would be necessary if priorities are set in terms of the relative funding or support for different industries, missions or fields of science.

Technological forecasting and economic forecasting is required if economic gains are to be a major component of science strategy. There is little point in developing a new technology for which subsequently there is no market because of changing economic conditions. Similarly, there is little point in developing a technology in a country when there are good reasons to believe that another technology will be developed in another country and supersede the indigenous technology, possibly even before it is developed.

In Japan's case, considerable attention is given to likely future demand for new technology and its products before priorities are determined. In estimating future demand, attention has to be paid to a wide range of factors including income trends, population trends and political changes which may alter world trading conditions. Progress of competing technology also has to be assessed.

There is some debate about the extent to which governments should engage in forecasting. It is claimed that so far as commercial developments are concerned government bodies are likely to be 'out of touch' and make poorer estimates than businessmen. On the other hand, government bodies may be able to take a wider perspective than businessmen and foresee difficulties which would not normally come to their attention. In cases where the government itself is directly spending on R & D, some assessment of broad costs and benefits would seem desirable and the assessment of these requires forecasting.

This is not the place to discuss different techniques of technological, economic and social forecasting.[25] A variety of techniques and models exist each with its own peculiar limitations. However, it may be pertinent to note that it is as a rule easier to predict the progress and application of new technology in its later stages than to forecast the likely results or net value of scientific effort. When new technology is at the pilot stage or has been commercially adopted for the first time, it may not be difficult to predict the rate of its diffusion or adoption. Very often long time lags are involved and a country which is aware of developments at the pilot or innovation stage and their likely economic potential can avoid being a laggard when this is in its economic interests.

Before becoming involved in detailed analysis, it may be useful by way of background to review broadly ideas about the role of science and technology in economic development, consider whether we should be optimistic about the ability of science to continue to contribute to economic development and look more critically at the need and problems inherent in government support for science and technology.

1.7 Critical views about the role of science and technology in economic development

Scientific and technological progress has in general been welcomed by economists and has been a source for optimism about the future of mankind. At least since Adam Smith[26] in the eighteenth century economists have stressed the important role of science and technology in raising living standards. When Malthus[27] pointed out at the end of the 1700s that tendencies of population to increase combined with diminishing returns in production might cause incomes to fall to subsistence level with economic growth, Ricardo[28] was quick to point out that this tendency could be staved off by technological progress.

Ricardo's view was eagerly taken up and embellished by Frederick Engels, Marx's friend and benefactor. It is worthwhile quoting Engels at length. He says:

> Yet, so as to deprive the universal fear of over-population of any possible

basis, let us once more return to the relationship of productive power to population. Malthus establishes a formula on which he bases his entire system: population is said to increase in a geometrical progression – 1 = 2 = 4 = 8 = 16 = 32 etc., the productive power of the land in an arithmetical progression – 1 = 2 = 3 = 4 = 5 = 6. The difference is obvious, is terrifying: but is it correct? Where has it been proved that the productivity of the land increases in an arithmetical progression. The extent of land is limited. All right! The labour-power to be employed on this land-surface increases with population. Let us even assume that the increase in yield due to increase in labour does not always rise in proportion to labour: there still remains a third element, which, of course, never means anything to the economist – science – whose progress is as unceasing and at least as rapid as population. What progress does the agriculture of the century owe to chemistry alone . . . But science increases at least as much as population. The latter increases in proportion to the size of the previous generation, science advances in proportion to the knowledge bequeathed to it by the previous generation, and thus under the most ordinary conditions also in geometrical progression. And what is impossible to science?[29]

It is interesting to note that Rescher in his recent book on *Scientific Progress* missed this important passage and accordingly was at a loss on the basis of Engels' *Dialectics of Nature* to understand the basis for current Soviet Marxist views on scientific progress. According to Rescher: 'Soviet writers tend to reject the idea that there are any limits or limitations to scientific progress. For it is felt that a limit on science entails a limit to technological progress . . . Soviet writers on scientific growth almost unanimously dismiss the picture of logistic development of science often favoured in the West.'[30]

While possibly not as optimistic as Engels, mainstream economists during the remainder of the nineteenth century and this century have continued to be optimistic about the contribution of science and technology to continuing economic development. This can be seen for example from Samuelson's introductory economic text.[31]

For most mainstream economists it has been claimed the doomsday philosophies like those of Meadows[32] linking the growth of science and economic growth to impending disaster for mankind has been but a ripple on the surface or a reason only for a marginal change in course. This has led one non-orthodox economist to quip that most mainstream economists are trying to find the optimal seating arrangement on the *Titanic*. In all fairness, a number of economists, for example E. J. Mishan[33] and Kenneth Boulding,[34] have expressed their doubts about the desirability of economic growth and the value of commonly held social and economic goals. Mishan specifically points out that new technology may not serve the needs of man. He says in *The Costs of Economic Growth*:

The younger generation will be facing the future with honesty only when it brings itself to face the strain of thinking through the consequences, tangible and intangible, certain and speculative, of the current drift into the future and, in doing so, recognizes that in the new world the old liberal harmonies are not to be found; that on many issues painful choices have to be made, and in some cases the needs of men and the needs of technology may prove to be irreconcilable.[35]

On the whole, established economists remain optimistic about progress in science as the means to reduce or avoid the possible harmful side-effects of economic growth such as increasing levels of pollution, environmental degradation and resource depletion. Science, however, may have to be channelled in the correct direction *possibly* by the government. Representatives of this optimistic view include Nordhaus[36] and Beckermann.[37] For example Nordhaus has argued persuasively that science and technology provides us with the ability to maintain and even increase our living standards despite the growing depletion of fossil fuels.

Empiricists also emphasize the role of science and technology in economic development. When Rostow[38] claimed in the 1950s that it was necessary for a nation to invest 10% or more of its net national product in order to begin on a path of sustained development and economic growth, this 'big push' doctrine was challenged. Blum, Cameron and Barnes pointed out that historical 'research indicates that almost every developed country of today entered a phase of sustained growth with investment ratios below the magic figure of 10%; and that the rise in that rate followed, rather than preceded, the adoption of new technologies.'[39] In Great Britain sustained economic growth began in the eighteenth century as Phyllis Deane[40] points out even though the investment ratio was below 5%. Economic growth occurred because new inventions were being embodied in the capital stock and education was improved. Economic development in France and Germany seems to have had a similar genesis. More recently, the applied economist Edward Denison estimated that almost a half of the growth in American GNP between 1929 and 1959 was due to increased education and improved technology.[41]

Whether or not less developed countries (LDCs) can repeat the pattern of European development is debatable. A number of writers believe that the fact that European countries and a few others have developed makes it more difficult for LDCs to make economic progress. In particular, those holding the centre-periphery theory of economic development argue that developed countries (the centre) dominate economic change in the periphery (LDCs) in such a way that scientific and technological breakthroughs in LDCs are extremely unlikely. Scientific and new technological change in LDCs is bound to be marginal and such countries are very dependent upon the import of foreign technology which may be inappropriate to their resource-mix. In the circumstances, it is argued that existing LDCs cannot repeat the pattern of European economic development and according to Marxists are exploited by the developed countries. In contrast neo-classical economic theory maintains that the greater world trade which can be expected to accompany economic growth in any part of the world is likely to be a powerful force for increasing incomes throughout the whole world and altering specialization in production by countries so that all gain. Development in any part of the world provides greater opportunities to LDCs to develop because incomes in these countries increase, so providing local funds for investment and more funds are available for foreign investment in LDCs. Neo-classical economic theory sees foreign investment (free international capital movements) as a means to raise incomes throughout the world whereas Marxists see this as a means of neo-capitalist exploitation. Thus, there are at least two competing theories of current economic development to take account of when considering scientific and technological change in LDCs.[42]

There are also other pessimistic views about the role of science and technology in economic development. They have, for example, been put forward by members of the neo-Marxist Frankfurt School and with modification appear to have been embraced by Hilary Rose and Steven Rose.[43] Broadly this School sees scientific and technological change under corporate capitalism and bureaucratic socialism as being geared to production and the domination of nature and man. Man is increasingly alienated and oppressed by technological change in capitalistic systems including bureaucratic socialist systems as in the Soviet Union. The Roses point out that one member of this group:

> Marcuse, in a century of massive growth in the scale and power of science, discerns science and technology as a particular mode of rationality aiding human oppression, either directly as the technology of repression, or individually through biological manipulation: 'Technology seems to institute new, more effective and more pleasant forms of social control and social cohesion.' Thus political questions are dissolved into technical issues to be resolved by experts. Technological rationality becomes political rationality.[44]

Other neo-Marxists, such as Ernst Mandel, emphasize that scientific and technological change becomes of increasing importance in late capitalism as a means for staving off an economy-wide decline in profits or surplus value.[45] The whole of capitalist apparatus including the instruments of government and its institutions, such as educational bodies including universities, is directed towards maintaining surplus value through developments in science and technology. In the process labour is exploited, degraded and alienated according to this view. Despite this pessimism, it seems clear that at least in developed countries, standards of living have risen markedly in the last two hundred years and on the whole the lot of the working class has improved. While Marx's prediction of the increasing immiserization of the working class under capitalism and with the passage of time has not, as yet, come to pass, alienation of participants in industrialized economies is a continuing problem.

1.7.1 Should we continue to place our hopes in scientific progress?

It can be convincingly argued that in the span of two centuries science and technology (but not alone) has enabled our standard of living to rise to a level which would have been hard to imagine in the 1700s and has enabled the globe to support a vastly increased population. No doubt Engels would have been impressed by the apparent geometrical progression of science and technology in this time span. Can we expect this progression to continue? Views are divided about this as Rescher points out in his recent book.

Some still remain as optimistic as Engels but there are others who believe that the possibilities for scientific progress will soon *terminate* because all that man can discover and assimilate will have been discovered or nearly so. Rescher's own position[46] is less pessimistic. He argues that scientific progress will not terminate but will decelerate. After a period of apparent increasing marginal returns from scientific effort we shall now, in his view, enter the phase of decreasing returns to scientific effort. He predicts that in the future, resources for scientific effort will remain stationary but the returns obtained from such effort will decline for the world as a whole. Are we coming to the end of the era of scientific discoveries? I must say that I do not know. I have not seen convincing evidence of an end to this era yet.

1.8 Critical views of government support for science and technology

Can the supply and application of science and technology (S & T) be safely left to the market system without government intervention and support? In my view there is reason to believe that free markets will undersupply S & T effort in many fields and misdirect its application from a social view, but of course the extent of the undersupply depends upon the rapidity with which marginal returns from S & T effort are declining. If one is numbered among the pessimists about the basic *possibilities* for future scientific progress, then government support for S & T effort will be of little or no avail and the only effective role for the government will be in ensuring the transmission and preservation of existing knowledge and regulating its social use.[47] I believe that there is no reason for such pessimism.

Individuals or companies can only appropriate a fraction of the social (marginal) gain from most of their S & T efforts and this fraction is likely to vary from project to project and industry to industry. Consequently, given the neo-classical economic view of the world in which firms try to maximize their profit, total S & T effort is likely to be less than socially optimal and the pattern of effort distorted unless government support is given to bring (marginal) private gains into line with (marginal) social gains. Favourable overspills will be greater in some industries – possibly those that are more purely competitive in nature with many small firms than is the case of monopolistic larger ones. Overspills in basic science are likely to be greater than in development and so on. On the other hand, negative overspills are also possible and government taxes or control may be required here, e.g. to check the use of an especially polluting technique.

Within the neo-classical system of economic thought and as summarized earlier, risks and uncertainties, imperfections in financial markets and the failure of the market system to direct the rise of whole new industries may also be reasons for government intervention in and support from science and technology.

It should also be pointed out, however, that a slightly different picture is painted by some theories of corporate capitalism (for example those of Galbraith and Marris) and by some New Left and Marxist thinkers.[48] Galbraith, for instance, following a point of view developed by the American economist, Veblen, suggests that there may be excessive spending on R & D and over rapid introduction of technology by corporations (especially large corporations) in the industrial sector. This is because managers follow their own goals rather than those of the shareholders of their companies and are preoccupied with maximizing their sales. Technocrats (the managerial, scientific and technical class) are the dominant group in present-day society and according to Galbraith have strong links with government and the government bureaucracy and movement between government and private bureaucracies is frequent. Industrial technocrats concentrate on developing new products and markets, manipulate demand and foster over-consumption. This view, however, would be consistent with the presence of underinvestment in R & D from a social viewpoint in industries consisting of small firms and in areas such as basic science.

There has been little testing of theories of corporate capitalism. A similar *relative* distortion of R & D effort can occur if corporations do maximize profit but the Galbraithian position is that this distortion is exaggerated under corporate capitalism. Whether it is exaggerated to such an extent as to cause excessive industrial S & T effort in the corporate large firm sector is a moot point.

1.8.1 Dangers of government support and intervention in S & T

While there may be a need for government support for science and technology, it does not follow that government intervention will necessarily benefit the community. Just as markets are subject to failure so is government because of imperfections in political and adminstrative mechanisms of decision-making.

The following political imperfections have been suggested by the work of authors such as Downs, Buchanan and Tullock, and Niskanen[49] but of course not everyone is convinced that their theories hold.

1. Government departments are in a symbiotic relationship with client groups. Large producers are the dominant and politically most active client group. This implies that they will have a more than proportionate influence on government policies in support of science and technology. There is a risk that Galbraithian biases are not eliminated but exaggerated in the democratic political system by government supports for S & T.

2. Politicians in order to obtain votes may concentrate government support on big science and areas where effort more easily comes to the eye of the voter.

3. Distortions in government support for S & T may arise from the supposed tendency of individual government departments to try to maximize their budgets or command over-resources.

4. Even if government departments and politicians altruistically support the social interest in implementing S & T policy

   (i) the difficulty of defining the social interest or a common goal exists, and
   (ii) co-ordination and informational barriers between and within departments as well as the finite capacity of individuals may prevent a common goal from being efficiently pursued.

5. Bureaucrats may be even more imperfect than company managers in their prediction of future events. Bureaucrats may not be skilled at 'picking winners'.

These possible dangers of government intervention in science and technological effort need to be offset against benefits before recommending greater government participation. However, we are faced with a choice between imperfect markets and imperfect political mechanisms. In some circumstances, an imperfect political mechanism of intervention will be socially superior to an imperfect market mechanism in guiding science and technological effort.

It is only by recognizing the limitations of political and administrative mechanisms that we can improve upon them. A useful step towards improving government science and technology policy is a better understanding of the possibilities for and limitations to such policy, a part of which is covered by the subject of technology assessment.[50] The analysis and review in the next two chapters of alternative strategies for science and technology policies and their implications should assist in the planning and assessment of actual policies.

Notes and references

1. In the following OECD countries, the government sector in 1973 performed the percentage of Gross Expenditure on Research and Development (GERD) indicated in brackets: Australia (40.6), Belgium (16.0), Canada (33.0), France (25.0), West Germany (15.4), Japan (13.2), Netherlands (19.9), Sweden (8.3), United Kingdom (25.7), United States (15.0). In general the proportion of GERD financed by the government sector was much higher. OECD (1976), *Science Resources Bulletin*, (1) April.
2. Hetman, F. (1973), *Society and the Assessment of Technology*, OECD, Paris, p. 7.
3. Green and Morphet define technology in a similar way. They define technology as the systematic knowledge of technique and technique as 'the interaction of person(s)/tool or machine/object which defines a way of doing a particular task.' Green, K. and Morphet, C. (1977), *Research and Technology as Economic Activities*, Butterworths, London, p. 2.
4. Feibleman, J. K. (1961), Pure science, applied science, technology, engineering: an attempt at definitions, *Technology and Culture*, 2 (4), pp. 305–17.
5. Nelson, R. R. (1959), The simple economics of basic scientific research, *Journal of Political Economy*, **67**, pp. 297–306.
6. For a review of the patent system, for example:
   (a) Firestone, O. J. (1971), *Economic Implications of Patents*, University of Ottawa Press, Ottawa.
   (b) Machlup, F. (1958), *The Production and Distribution of Knowledge in the United States*, Princeton University Press, Princeton.
   (c) Taylor, C. T. and Silbertson, Z. A. (1973), *The Economic Effects of the Patent System*, Cambridge University Press, Cambridge.
7. (a) Arrow, K. J. (1962), Economic welfare and allocation of resources to invention, in *Rate and Direction of Economic Activity*, National Bureau of Economic Research, Princeton University Press, Princeton.
   (b) Arrow, K. J. and Lind, R. (1970), Uncertainty and the evaluation of public investments, *American Economic Review*, **60**, pp. 364–78.
8. For a further discussion of risk overspills and illustrations of how technological risks can result in socially inferior behaviour in our society see:
   (a) Tisdell, C. (1979), Limitations on the social and economic use of the environment, in: *The Status of the National Environment*, Institute of Engineers, Canberra, pp. 22–27.
   (b) Tisdell, C. (1979), Scientific and technological risk-taking and public policy, in *Theory of Knowledge and Science Policy* (eds W. Callebaut *et al.*), University of Ghent, Belgium, pp. 576–85.
   (c) Tisdell, C. (1980), *Law, Economics and Risk-Taking*, RROP No. 57, Department of Economics, University of Newcastle, October.
9. Pavitt, K. (1976), Government support for industrial research and development in France: theory and practice, *Minerva*, **14**, pp. 331–54.

10 See for instance: Kahn, A. E. (1966), The tyranny of small decisions: market failures, imperfections and the limits of economics, *Kyklos*, **19,** pp. 23–47.
11 Marshall, A. (1925), *Principles of Economics*, 8th edn, Macmillan, London, pp. 266, 314, 441.
12 (a) Neutze, G. M. (1965), *Economic Policy and the Size of Cities*, Australian National University Press, Canberra.
 (b) Tisdell, C. (1975), The theory of optimal city-sizes: elementary speculations about analysis and policy, *Urban Studies*, **12,** pp. 61–70.
13 Oshima, K. (1973), Research and development and economic growth in Japan, in: *Science and Technology in Economic Growth* (ed. B. R. Williams), Macmillan, London, pp. 310–23.
14 Pavitt, K. (see ref. 9). Note that Pavitt and his co-author Walker, take a rather different tack in Pavitt, K. and Walker, W. (1976), Government policies towards industrial innovation: a review, *Research Policy*, **5,** pp. 11–97. They argue strongly that *empirical* information (both industry-specific and firm-specific) is needed about *why* innovations are (and are not) being made if one is going to formulate relevant government policy to remove the hindrances to economically and socially desirable innovations.
15 Organisation for Economic Cooperation and Development (1975), *Changing Priorities for Government R & D*, OECD, Paris, p. 77.
16 OECD (see ref. 15).
17 Kuznets, S. (1974), *Population, Capital and Growth: Selected Essays*, Heinemann, London.
18 (a) Organisation for Economic Cooperation and Development (1971), *Science, Growth and Society*, OECD, Paris.
 (b) Meadows, D. et al. (1972), *The Limits to Growth*, Earth Island, London.
 (c) Mishan, E. J. (1967), *The Costs of Economic Growth*, Staples Press, London.
19 Nordhaus, W. and Tobin, J. (1970), *Is Growth Obsolete?* Cowles Foundation Discussion Paper No. 319, December, Yale University, New Haven.
20 Beckermann, W. (1973), Economic growth and welfare, *Minerva*, **11,** pp. 495–515.
21 Beckermann, W. (see ref. 20), p. 513.
22 (a) Pavitt, K. (see ref. 9).
 (b) Smith, B. R. (1973), A new science policy in the United States, *Minerva*, **11,** pp. 162–74.
23 (a) His approach involves an organizational view of behaviour. Compare for instance, Cohen, K. and Cyert, R. M. (1965), *Theory of the Firm*, Prentice-Hall, Englewood Cliffs, Ch. 16 and 17, and see Lindblom, C. E. (1958), Policy analysis, *American Economic Review*, **48,** pp. 298–312.
 (b) Lindblom, C. E. (1959), 'The science of muddling through', *Public Administration Review*, **19,** pp. 79–88.
 (c) Hirschman, A. O. and Lindblom, C. E. (1962), Economic development, research and development, policy making: some converging views, *Behavioural Science*, **7,** pp. 211–22.
 (d) Lindblom, C. E. (1977), *Politics and Markets*, Basic Books, New York.
24 Hirschmann, A. O. and Lindblom, C. E. (see ref. 23c). Reprinted in Emery, F. E. (ed.), (1969), *Systems Thinking*, Penguin, Harmondsworth, pp. 357–8.

25 (a) Aujac, H. (1973), A new approach to technological forecasting, in: *Science and Technology in Economic Growth,* (ed. B. Williams), Macmillan, London, pp. 96–115.
    (b) Bright, J. and Schoeman, M. (1973), *A Guide to Practical Technological Forecasting,* Prentice-Hall, Englewood Cliffs.
    (c) Byatt, I. and Cohen, A. (1969), *An Attempt to Quantify the Economic Benefits of Scientific Research,* Science Policy Studies No. 4, HMSO, London.
    (d) OECD (1967), *Technological Forecasting in Perspective,* OECD, Paris.
26 Smith, Adam (1776), *Wealth of Nations.*
27 Malthus, T. R. (1798), *An Essay on the Principle of Population as Affects the future improvements of Mankind.*
28 Ricardo, D. (1817), *The Principles of Political Economy and Taxation,* 1st edn; (1821), 3rd edn.
29 Engels, Frederick (1959), Outlines of a critique of political economy, in: *Economic and Philosophic Manuscripts of 1844* (K. Marx), Foreign Languages Publishing House, Moscow, p. 204.
30 Rescher, N. (1978), *Scientific Progress,* Basil Blackwell, Oxford, pp. 124–5.
31 Samuelson, P. (1970), *Economics,* (8th edn), McGraw-Hill, New York, Ch. 37.
32 Meadows, D. H. *et al.* (1972), *The Limits of Growth,* Universe Books, New York.
33 Mishan, E. J. (1967), *The Costs of Economic Growth,* Staples Press, London.
34 Boulding, K. (1970), *Economics as Science,* McGraw-Hill, New York. For a controversial review of economists on economic growth, see Daly, H. E. (1979), Entropy, growth and political economy of scarcity, in: *Scarcity and Growth Reconsidered* (ed. V. Kerry Smith), Resources for the Future, Baltimore, pp. 67–94.
35 Mishan, E. J. (see ref. 33), pp. ix, x.
36 Nordhaus, W. D. (1974), Resources as a constraint on growth, *American Economic Review Papers and Proceedings,* pp. 22–26.
37 Beckermann, W. (see ref. 20).
38 Rostow, W. W. (1952), *Process of Economic Growth,* Norton, New York.
39 Blum, J. *et al.* (1967), *The Emergence of the European World,* Routledge and Kegan Paul, London.
40 Deane, P. (1955), The implications of early national income estimates for the measurements of long-term economic growth in the United Kingdom, *Economic Development and Cultural Change,* **4,** (1), pp. 3–38.
41 Denison, E. F. (1962), *Sources of Economic Growth and the Alternatives before Us,* Committee for Economic Development, New York.
42 For a review of some of the relevant theories of economic development and further references, see, for example:
    (a) Tisdell, C. (1977), Imperialism and traditional economic view of development, *Research Report or Occasional Paper,* No. 38, Department of Economics, University of Newcastle, Australia.
    (b) Resnick, S. A. (1975), State of development economics, *American Economic Review,* **65,** (2), May, pp. 317–32.
    (c) Myrdal, G. (1956), *On International Economy: Problems and Prospects,* Routledge and Kegan Paul, London.
43 Rose, Hilary, and Rose, Steven (eds) (1976), *The Political Economy of Science,*

The Macmillan Press, London.
44 (see ref. 43), pp. 25–6.
45 Mandel, Ernst (1975), *Late Capitalism*, NLB, London, especially Ch. 8.
46 Rescher, N. (see ref. 30). Attention should be brought to another pessimistic view about the prospects of future technological progress that has a different basis to Rescher's. Joseph Schumpeter, while pointing out that capitalism has been an important force for innovation and scientific progress (for example through the process of competition between companies by means of new products and new processes), saw a danger of corporations becoming larger with economic development. Scientific and technological effort would consequently in his view become more institutionalized (more often carried out in larger corporations and organizations), so stifling imaginative researchers and reducing the rate of technological progress. His predicted slow-down in scientific progress has an organizational basis. See Schumpeter, J. A. (1942), *Capitalism, Socialism and Democracy*, 2nd edn, Harper Brothers, New York.
47 The pessimistic and the more optimistic position about the value of government support for scientific and technological effort given the possibilities for technological progress can be illustrated by Fig. 1.3 below. P_1P_1 represents the private marginal return from investment in S & T and S_1S_1 the social marginal return under a pessimistic hypothesis of rapidly decreasing returns. Private interest leads to an investment of K_1 in S & T effort. The socially optimal level of investment is K_2. Private effort is below the socially optimal level by $K_2 - K_1$. But only a small social gain is made by government support for S & T. A subsidy of v on *increased* private effort would imply public support of $v(K_2 - K_1)$ for S & T effort. But if decreasing returns are much less rapid and social and private marginal returns like those indicated by S_2S_2 and P_2P_2 apply, greater benefits are achieved from public support and it is optimal to commit

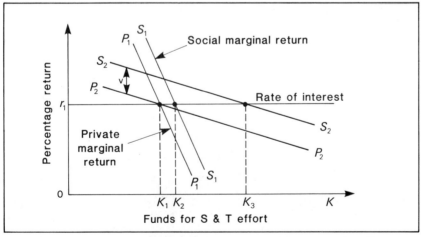

Fig. 1.3 The social value and effectiveness of public support for efforts to advance science and technology. These are dependent upon the inherent marginal returns to S & T effort.

greater funds for this purpose. In this case, the socially optimal amount of effort is K_3. An expansion from the private level of K_1 to the socially otpimal level of K_3 could be achieved by a subsidy of v on additional S & T investment. This would involve a public outlay of $v(K_3 - K_1)$. The value of public support for S & T effort depends upon the marginal productivity of that effort and ultimately as suggested by Rescher this may be governed by natural laws which governments cannot change, just as Canute could not get the tide to respond to his order to recede.

48 For a summary of Galbraith's views see Gwartney, J. D. (1979), *Microeconomics*, Academic Press, London, Ch. 10 and 11; and for a further discussion of these Hartley, K. and Tisdell, C. (1981), *Microeconomic Policy*, Wiley (UK), Cheshire, (to be published), Ch. 9 and 5. For a relatively recent Marxist interpretation of the role of science and technology in capitalism, see for example Mandel, Ernst (1975), *Late Capitalism*, NLB, London, especially Ch. 8.

49 See, for example:
(a) Downs, A. (1957), *An Economic Theory of Democracy*, Harper and Row, New York.
(b) Buchanan, J. and Tullock, G. (1962), *The Calculus of Consent*, University of Michigan, Ann Arbor.
(c) Niskanen, W. A. (1971), *Bureaucracy and Representative Government*, Aldine, Chicago; and a recent Australian study by Anderson, Kym (1980), The political market for government assistance to Australian manufacturing industries, *The Economic Record*, **56**, (June), pp. 132–44.

50 For a review of some approaches to this subject see Medford, D. (1973), *Environmental Harassment or Technology Assessment*, Elsevier, Amsterdam, and Hetman, F. (ref. 2).

CHAPTER TWO

Science Policy Options and Priorities

2.1 Introduction

In its widest application *science policy* is concerned with education, the stock of knowledge, its availability and use, and research and development. *Technology policy* is concerned with the adoption and use of techniques – innovation, diffusion of techniques and their replacement. As is indicated in Table 2.1, however, the borderline between the two policy types is not clearcut.

Division between the areas and variables of science policy and technology policy are not watertight. Education and the stock of knowledge, for instance, play an important role in influencing the rate of innovation and diffusion of new technology. Again, all elements in the left-hand column of Table 2.1 influence those in the right-hand column and there is interdependence between variables in the left-hand column.

This chapter concentrates on the areas of science policy listed in the first column of Table 2.1 and the next chapter on technology policy reviews

Table 2.1 Spheres of science policy and technology policy.

Area of science policy	Area of technology policy
'Fuzzy' separation interdependence between variables	
Education	Adoption and application of techniques
Stock of knowledge – its application and availability	– innovation
	– diffusion of techniques
Research and development	– replacement of techniques or equipment

those areas listed in the second column. The aim is to cover these areas broadly.

Some of the questions considered in this chapter are: How is education and the stock of knowledge of significance for scientific and technological progress? What role can educational institutions play in promoting technical progress? Why is it efficient for higher educational institutions to carry out teaching jointly with research? What important decisions must a society make about preserving, transmitting and making available its stock of knowledge? Why are market mechanisms imperfect for knowledge transmission and preservation? Other questions considered under the heading of research and development activity include: Why may it be beneficial to co-ordinate science policy with industrial policy? To what extent should a nation depend on imported scientific knowledge rather than produce scientific knowledge at home? What role can government science policy play as an instrument of social policy? In particular, what case exists for government support of scientific effort designed to improve the environment and health and expand energy prospects? What particular problems arise and what alternatives are foregone as a result of government expenditure on defence research and development and big science? What national balance is desirable between basic research, applied research and development? Is it most advantageous for a nation to concentrate or specialize in a few areas of scientific enquiry or should it cast its net wide in its scientific endeavours? Should scientific activity be geographically concentrated within a country? Does the emergence of new fields of scientific enquiry pose special problems for government science policy? Is there reason to suppose that inertia results in undue concentration of national resources in established fields of scientific enquiry to the relative neglect of emerging fields? Other aspects considered are: How should R & D performance be distributed between those able to undertake R & D? In particular, what relative role should be played by the government, private industry, universities and independent non-profit research institutions in the performance of R & D? Should domestically owned firms or bodies be preferred to foreign-owned firms or bodies in the performance of R & D? What is service science and what government assistance is needed with it? What role does science play in international affairs and what scope exists for co-operation and competition between nations in scientific effort? Is foreign science and scientific aid a help or a hindrance to the economic development of less developed countries?

The above may appear to be a formidable list of questions but they are all questions which cannot be ignored in formulating government science policy.

2.2 Education and the stock of knowledge

Education (largely but not exclusively) involves the transmission of systematic and formulated knowledge and therefore is a scientific-type activity. Knowledge transmitted forms a basis for scientific and technological advance and a basis for practical use and helps pass on the culture and values of a society. The stock of knowledge is an important building block for education and a source for direct application. It therefore needs to be given special attention.

Figure 2.1 sets out some of the main links between variables to be taken into account in scientific and technology policy. It indicates that education is in most cases an important prerequisite for extending knowledge. It enables potential researchers to come quickly to the boundaries of their fields and not duplicate earlier work, and provides them with tools or components in the form of concepts and ideas for creative thinking.

It should be observed that Fig. 2.1 does not represent a complete system. Some feedbacks and interrelationships are not shown. For instance, scientific advances feeds back to increase the stock of knowledge. Furthermore, the advance of science is influenced by the transmission of values

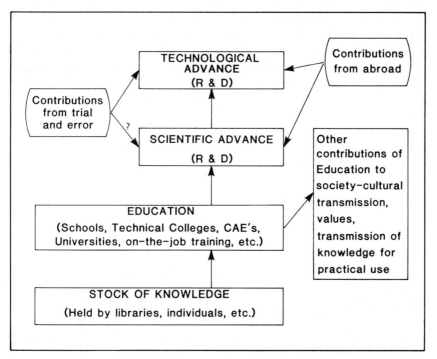

Fig. 2.1 Interrelationships between main variables to be taken into account in science and technology policy.

and attitudes to research and enquiry. Appropriate values can be a positive force making for scientific advance.

2.2.1 Education

Education is an important factor in the supply of science and technology. Carter and Williams have stressed this in the following way:

> Our supply of people capable of using science, and of adding to it by research, depends on education. Likewise our supply of people capable of using technology and of adding to it by design and development work depends on education. Research and development grow out of education, development grows out of research, innovation (the use of an invention to create new products and processes) grows out of development.[1]

Apart, however, from teaching students, some of whom will later contribute to scientific and technological advance, higher educational bodies undertake research directly. The amount of resources and effort devoted to research varies between institutions.

The proportion is typically high for universities and low for technical colleges. The research emphasis of universities is on basic science. Measured by their total expenditure of GERD, the following proportions of total research and development activity in brackets were attributed by the OECD to higher educational institutions in the following countries in 1973: Australia (17.2), Belgium (26.2), Canada (31.0), France (14.4), Germany (19.7), Japan (18.1), Netherlands (19.6), Sweden (24.6), United Kingdom (8.8) and United States (12.0). These are not insignificant proportions. Science policy requires decisions about the proportion of resources to allocate to higher educational bodies for research and teaching, how to allocate funds to projects and to individuals and how to apportion funds between the different institutions. There are likely to be no 'magical' desirable percentages for such allocations.

It might be observed that a number of factors makes it efficient for universities and other tertiary educational bodies to supply education and research results *jointly*. The costs of pursuing these activities independently can be higher for the same level of output from the activities. This is because teaching, at the advanced level at least, and various research activities require similar overheads which can be used jointly. Libraries provide one obvious example. Very often, however, equipment and buildings can be jointly used for research and teaching. The productivity of staff in teaching may also be raised if they have job variety through research. Lecturers may need to engage in research if they are to understand certain scientific developments either at home or abroad and teach them. Active research may also be required to transmit techniques of research and enquiry.

Educational institutions can assist with the introduction and diffusion of new techniques into an economy. For many future employees they provide the point of first contact with techniques which they will employ in their workplace. This includes experience with scientific testing equipment, computers, typing and copying equipment and so on. To the extent that educational bodies lead rather than lag industry in the use of new techniques, they act as catalysts for economic change. If they lag substantially behind industry in equipment, as sometimes is the case, they place an increased burden on industry since industry must then provide increased on-the-job training. On the other hand, if training bodies scrap earlier techniques too far in advance of their replacement by industry, their training programmes can prove also to be of limited relevance to the workplace.[2]

Educational bodies and associated research institutions can often play a useful role in building up a core of expertise for a new industry before the industry becomes commercially viable. The development of the industry may depend upon the prior availability of such expertise. Industries based upon or associated with nuclear energy provide an example. Their development requires the existence of individuals with expertise and knowledge in the nuclear field. It might be noted that Japan used research institutes to great advantage after the Meiji Restoration to introduce new ideas to Japan and build up a body of expertise in selected scientific areas of economic value.[3]

In setting priorities, consideration also needs to be given to other interrelationships between the educational sector and the rest of the economy. To what extent should the specialized equipment and expertise of universities and related bodies be made available to industry, government and other interests? To what extent should the research activities of universities and related bodies be directed towards filling community needs as perceived by those able to make research funding available? To what extent must higher educational institutions be independently funded or funded without strings attached so as to preserve freedom of enquiry? To what extent should a country rely upon a system of tied research grants to finance the major part of research in universities? These are difficult questions to answer.

The interrelationship between university research in Japan, for example, and economic needs does not appear to be clearcut. On one hand Oshima, for example, says:

> In most cases there is little direct link between basic research in universities and public institutions on the one hand and industry on the other. And the contribution of activity in basic research has been mainly to supply a scientific and technological soil of high standard in which to cultivate active technological innovation in industry.[4]

36 *Science and Technology Policy*

Later he says:

> There was comparatively little effort devoted to national research and development projects for non-economic purposes. The major research effort was directed to economic applications. This was equally true of basic research in universities.[5]

Consideration needs to be given to whether the composition of graduates by fields from universities and higher educational institutions is likely to dovetail with future employment demands in the economy. Should this matter be left to free choice and (delayed) market feedback or should some manpower planning or prediction of future requirements be made and this information be supplied to prospective students? Should scholarships be varied in value or in available number by fields to encourage desired changes in the composition of graduates?

2.2.2 *Stock of knowledge*

Existing knowledge and new knowledge need to be recorded, stored and made available for future use if the cultural and economic progress of man is not to be impeded. In general, the technical progress of societies with inability or little ability to record knowledge, store it and make it widely available, tends to be slow.

There are many policy questions and priorities to consider as far as the stock of knowledge is concerned. Does the social and economic system encourage the recording of new knowledge once it is obtained? Are records of existing or previous knowledge adequately maintained? Are records available to a sufficiently wide audience and without unreasonable delay?

The recording of knowledge is likely to depend on the rewards for this and the cost of recording it. Rewards may take the form of recognition by learned societies and other researchers, accelerated promotion and royalties from publications stemming from copyright.

Copyright of publications, royalties and the use of the market profit mechanism by publishers is unlikely to result in the recording and adequate availability of all knowledge which is worthwhile recording and making available from a social point of view. These mechanisms by themselves are likely to result in less recording and availability than is socially optimal. This is because (a) those recording information and making it available cannot appropriate all the benefits at its margin of use and (b) copyrights like the patent system give at least temporary monopolies which in some cases may be exploited by inflating prices for publications so reducing the volume of sale of the publication and the diffusion of knowledge. One way to correct for these deficiencies is to provide a suitable subsidy to publication. This subsidy could be on the volume of the published item, e.g. number of copies of a book published.

Some publications, however, have such low prospects of substantial sales that the costs of producing them cannot be commercially recovered at all, even using copyright regulations. However, this does not imply that the benefits to society of such a publication would be less than the costs of its production. Commercial receipts do not capture all gain from a publication. For instance, a publication may lead to the discovery of further knowledge of great value (stemming knowledge) not reflected in the payment for the original publication. Furthermore, the information in a publication can be communicated without payment or with minor payment to the supplier of this information, for instance by word of mouth or by the use of books in public libraries. Publications for which there is a low volume of demand but have a positive net social value may need to be assisted financially by the government. This could be done by providing financial assistance for the preparation of manuscripts (typing and other costs) and subsidizing publishers of manuscripts in this class, such as university presses. There are, however, some thorny issues involved in selecting manuscripts or prospective manuscripts for such assistance.

Libraries are the most important repositories for information in our society. Unless adequate funds are provided for libraries, the available stock of information in a country may be inadequate for scientific demands and may hamper social and economic progress. Libraries use resources and must compete with other uses for these resources. Consequently, libraries generally are unable to obtain sufficient funds to store and preserve all available knowledge and information. Priorities have to be established for the level of funding for libraries and the way in which libraries use their limited funds.

Continuing attention needs to be given to least cost methods of storing material and to ensuring that information can be easily retrieved and made available. At the same time, consideration needs to be given to means of preventing physical deterioration of stored material. Inter-library loan systems may allow duplication of volumes to be reduced. While efficient inter-library loan systems may enable a library system to supply more information overall, against this must be set delays and the loss of browsing possible when publications are at hand. Similar difficulties occur with centralization. Even though national centralization of library facilities may reduce costs, it increases the risk of the national collection being wiped out by a catastrophic event.

The patent system provides an interesting example of a mechanism to promote flows of information in society. Patents provide an inventor with a temporary monopoly on the use of his invention in return for his making the knowledge of his invention available to society as a whole. The information may then be used by others to develop new inventions. However, the value of the system as a provider of information depends upon how well the invention is specified in the patent application, lags in processing such applications and the efficiency with which the patent office can retrieve specifications and make these available to others. In patent policy, and in science policy generally, priorities need to be established about how much information to store, what type of information to store and how quickly and efficiently to make it available.

2.3 **Research and development – general issues**

Many discussions of science policy concentrate almost exclusively on research and development activity. But R & D is only one component in the science sphere. The importance of R & D can easily be overrated or underrated. In this regard, Carter and Williams have observed that:

> ... discussions of the role of science in economic growth often turn into discussions of the role of 'R.D.'. Yet research and development are simply concerned with extending or improving our knowledge of science and technology. Current economic growth must be more dependent on success in applying *existing* science and technology than it is on current rates of expenditure on research and development. Current growth is related more closely to the stock of knowledge than the flow of new knowledge.[6]

Available evidence indicates that considerable returns can be obtained from research and development activity[7] but it by no means follows:

(a) that growth of GDP (Gross Domestic Product) is necessarily raised by increasing spending on R & D;
(b) that to increase R & D expenditure is the best use of resources;
(c) that the growth and profitability of industries with low levels of R & D intensity is improved, economically or otherwise, by greater R & D spending;
(d) that there is an optimal percentage of GDP to spend on R & D which can be determined *a priori* and is the same for all countries.

Spending on R & D need not increase growth since the resources allocated to R & D may give higher returns in alternative uses, e.g. in developing new mines or education or in improving management. The cost and benefits of alternative uses of resources needs to be carefully assessed in determining their optimal allocation. The nature of the R & D also determines the extent to which it results in economic progress. Research for defence purposes adds little to economic growth unless there is 'spin-off'. This is not to deny that increased R & D activity can be desirable and have useful growth and other effects but each case must be considered on its merits.

In particular, it cannot be assumed that the same percentage of GERD to GDP is optimal for all countries. Scarcity of resources and alternative opportunities for investment vary between countries and *ability to appropriate gains* from R & D may vary according to the size and state of development of a country. For instance Yoram Ben-Porath has argued that the smaller a country is and the less developed it is, the lower is the optimal level of its expenditure on R & D.[8] Ben-Porath contends that:

> If there were an effective way in which a country could trade the output of its R & D industry internationally, a small country could sell the output of its research industry and collect the bulk of the returns. But it is precisely the nature of R & D output that it is difficult to establish property rights in it, particularly internationally. The patent system is far from perfect and cannot be fully applied to all R & D output, particularly not to basic research. It is thus clear that smaller returns will accrue to a small country on a given R & D output.
>
> Large countries are likely to have bigger R & D programs in absolute terms. A large R & D program may have advantages over a small one because of the positive interaction between different R & D outputs. Variance in the probability of success tends to be smaller on a larger program composed of many projects, so that considerations of uncertainty also tend to favor large programs.[9]

While there is some substance in Ben-Porath's points the results or conclusion might not follow for all small countries. The conclusions may have to be modified for small countries such as Belgium, Sweden and Switzerland which are located in large markets and are home to some large multinational companies.

The most desirable balance between basic research and applied R & D is another bone of contention, and a case in which it is impossible to prescribe the optimal balance on an *a priori* basis. Smaller economies and less industrialized ones tend to concentrate relatively more than others on the basic end of the spectrum. This may be of assistance to them in the selection of technology from overseas and may help to keep their options open. But in the end, research must be pushed through to the development and technology stage and used if it is to contribute to economic growth or economic welfare. As Carter and Williams have said: 'It is easy to *impede* growth by excessive research, by having too high a percentage of scientific manpower engaged in adding to the stock of knowledge and too small a percentage engaged in using it.'[10]

Differences of opinion exist about the extent to which a government should financially support the development of new products. On the whole, it is argued that proportionately less government support is justified for the development of specific products than for basic research. For instance, Eads and Nelson claim:

> Economists have not always recognized adequately that the current regime of institutions and policies is unlikely to generate enough or the right allocation

of pioneering technological work. Much of the thinking about research and development implicitly draws a sharp distinction between basic research and product development – the former viewed as creating diffuse general-purpose knowledge and hence requiring public support; the latter, a patentable product whose profit potential adequately signals its social value. Yet much of R & D falls in between these extremes.

One important kind of R & D is research aimed at placing a particular technology on a stronger scientific footing. Another is experimental development to test the broad attributes of new product and process designs.[11]

On the other hand, they argue that government support for the development of specific civilian products such as SST (supersonic transport) or nuclear energy plants is liable to be misguided and divorced from market considerations.

The value of a new development depends upon its use and to a large extent the marketability of it or its products. This in turn depends upon the ability of management and its 'thrust' and whether or not there is adequate feedback between R & D teams and management in identifying production and market needs. Economic considerations and feedback to scientists about economic requirements are important considerations in developing new technology. This is underlined by Japanese experience of which Oshima has said:

> ... after success in achieving a high rate of economic growth as a result of technological innovation, it has become clear that a close co-ordination of technological and economic policies has been the major factor in successful technological innovation.[12]

2.4 Science and industrial policy

Scientific effort intended to support industry competes to some extent for resources with scientific effort for other purposes such as defence, the satisfaction of curiosity and the improvement of social welfare in a broad sense. Consideration needs to be given to the desirable degree of emphasis on these different purposes. In particular, the balance between scientific effort for economic purposes and for non-economic purposes needs to be considered carefully. The balance between these varies between countries. In Japan, the main emphasis is on R & D for economic purposes whereas in the USA the proportion of R & D funds for non-economic purposes is higher because of defence requirements and curiosity-type spending on the space programme. However, it might be observed that a proportion of the benefits of research for non-economic purposes can be in the form of spin-offs to industry. The space programme in the USA for instance accelerated the commercial development of mini-computers and miniature solar energy devices.

2.4.1 *Benefits from science for industry*

Research and scientific effort intended to assist industry can have a number of benefits. The following may benefit (when this effort results in appropriate technology):

(a) Industrialists through greater profits (which may be used to raise investment).
(b) Labourers as a result of improved conditions of work.
(c) Consumers by (i) lower prices for products, (ii) through the availability of better quality products incorporating additional characteristics or features and (iii) by the provision of a greater variety of products.

New technology resulting from scientific research for industry may improve a country's international trading position in the following ways:

(a) The value of its exports may rise as a result of the sale or increased sale of products produced by the new technology.
(b) The technology may be sold to overseas countries and royalties and licence fees would then add to export receipts.
(c) Companies, the major shareholding of which is held in the domestic economy, may invest directly overseas and increased dividends from abroad may be repatriated as a result of greater profit being earned by overseas subsidiaries. This results in greater domestic income and on improved balance of payments.
(d) The new technology may permit import replacement to occur and thus reduce the import bill.

Scientific and technological advance of relevance to industry can improve a country's balance of international payments and can also increase its gains from trade.

It is, however, appropriate to observe (a) that neither a large and continuing surplus nor deficit on the balance of payments is usually in a country's interest, and (b) in certain special circumstances, technological advance can reduce a country's gains from trade.[13] The latter can happen for instance when the demand for an export is inelastic abroad and the technology leads to cost reductions. The export price may fall by the whole of the cost reduction achieved. Special measures may need to be taken to ensure that a greater proportion of the world-wide gains from the advance is appropriated by the country in which the technological advance takes place.[14]

Traditionally, economists have argued that given world demands for different products, the specialization of countries in production and export of products depends upon their comparative costs of production of these commodities. In turn, comparative costs of production depend on factor endowments of countries, especially relative factor endowments. This implies that a country with relatively more land is likely to have a comparative advantage in products requiring the use of land, those with relatively more capital a comparative advantage in capital-intensive industries and so on. It is often further argued that in order to maximize its gains from trade, a country should specialize in the production of those goods in which it has a comparative advantage and that scientific efforts should be exerted to improve this advantage.

2.4.2 Product-cycle approach

The traditional point of view is being questioned from a number of angles:

(a) Comparative advantage in an industry may be 'manufactured' by a country if it obtains a lead in research, knowledge and experience in the field compared to other countries.
(b) Studies using labour and capital as the main resources, indicate that the USA does not specialize in the export of capital-intensive products but in the export of research-intensive products many of which are labour-intensive.
(c) The theory ignores the importance or possible importance of market power. It assumes that competition is perfect or pure.
(d) It ignores time-lags in which a country can establish a lead in a field and in which its firms can attain a temporary world monopoly or above normal profits before firms in other countries enter the field. Major national gains from trade can be made by being a scientific and technological leader in selected fields.

In discussing this new point of view,[15] Gruber, Mehta, and Vernon say:

> From capital and labour cost considerations, therefore attention has turned to questions of innovation, of scale, of leads and lags (Posner, 1961; Freeman, 1963, 1965; Hirsch, 1965; Hufbauer, 1965; Wells, 1966). [15] Approaches of this sort have tended to stress the possibility that the United States may base its strength in the export of manufactured goods upon monopoly advantages, stemming in the first instance out of a strong propensity to develop new products or new cost-saving processes. This propensity has usually been credited either to the demand conditions that confront the American entrepreneur or to the scale and structure of enterprise in U.S. markets. In any case, the propensity has given American producers a temporary advantage which has been protected for a time either by patents or by secrecy. Eventually, the monopoly advantage has been eroded; but by that time, the US producers have seized the advantage in other products.[16]

A typical product cycle for a firm or country innovating (and producing a successful product) might be like that shown in Fig. 2.2. The value of sales at first rises and then after an interval of time declines, as a result of entry and competition of new firms and countries into production of the product and/or as a result of the introduction of new superior products. Profits from the product innovation follow a similar course except that in the initial stages of the innovation the innovator (or innovating country) is likely to suffer a loss as he (it) spends to establish the product in the market and sorts out any production problems. His (its) profits may also continue to rise for a time after his (its) total sales revenue falls because experience continues to be gained with the production and distribution of the product. This may lead to a continuing reduction in costs with the passage of time. In the growth stage of the innovator's sales in the market, his profits tend to be high but are eventually eroded as new firms (or countries) commence production of the commodity and substitute products are developed. The level of profits achieved by a company and the gains from international trade made by a country are likely to depend upon its continuing ability to introduce or adopt new products early and successfully.

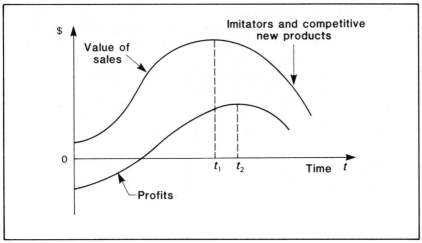

Fig. 2.2 Typical sales and profit profiles for a successful new product. The curves can be interpreted to apply to an innovating country or an innovating firm.

As Vernon has observed, the product cycle may give rise to typical patterns in international trade and investments.[17] New products are likely to emerge in a more developed country, such as the USA, which first of all caters for the home market and some small amount of export sales. After a time, export sales may rise and US companies may have sufficient volume of sales in other developed countries for economies of scale in production in these countries and may establish production plants there. Eventually as the US market declines, USA needs might be met by imports. At the next stage production gravitates to less developed countries, as demand falls in the other developed countries and in the end production is confined to LDCs.

2.4.3 *Profitability of being a leader or imitator*

An important factor to consider even if one accepts the product cycle approach is whether it is more profitable to be an innovator or a 'fast second', that is a speedy imitator of successful innovations. One may be able to modify a new product by 'reverse engineering' to avoid patent infringements if necessary. Baldwin and Childs argue that it is sometimes more profitable to the fast second imitator than an innovator.[18] They suggest that:

> Cost differences may favor the fast second imitator. It may be much cheaper and quicker to examine a competitor's product and ascertain its composition than it is to perform the research leading to the initial creation of such a product in the laboratory. Designing and implementing production processes for an item which has already been produced on a commercial scale by another may be somewhat less costly, possibly faster if the imitator possesses sufficient information to learn from his rival's mistakes, and undoubtedly far less uncertain than it is to carry out the original development. Thus reverse engineering, where feasible, is normally cheaper, quicker and less risky than innovation. Marketing and promotion costs may be reduced since the imitator may find it unnecessary to engage in testing of performance characteristics and side effects of the product in uses to which it might be put by various customers, and the imitator may benefit in both time and cost from customers' prior acceptance of the innovator's product.[19]

They consider, however, that where (after some lag) *many* imitators (including the fast rival) are likely to commence producing the product almost simultaneously, it is likely to be more profitable to be an innovator rather than a fast rival and to alter one's R & D effort accordingly. The expected (differential) speed with which competitors can be anticipated to commence supplying a new product is important in deciding whether it is more profitable to be an innovator or a fast imitator, or whether it is profitable for the firm or country to supply the product at all.[20]

If funding or support for industrial R & D is to be distributed selectively, it is important to take into consideration some of the factors mentioned by Baldwin and Childs. Other factors may also need to be taken into account in a product-cycle approach such as the international market power possessed by firms from the domestic country. Larger firms with international market power are likely to be able to appropriate a greater amount of the gains from an innovation than smaller firms with little market power. This can *some-*

times mean that greater national gains from exports can be made by supporting the R & D of large firms with market power than *small firms*.

2.4.4 More on product-cycle type approaches

If a product-cycle approach to technological gains is adopted a country may still find it worthwhile to take into account the comparative advantages which it has in view of its natural and other resource endowments. Other things equal, a country's gains from innovation may be greater in international trade if it chooses new products for which it is likely to have a comparative advantage in view of its resource endowments. In such cases, other countries are likely to exhibit longer imitation lags and this gives the innovating country a longer period in which to earn above normal returns. Of course, other elements in market failure such as industry-wide external economies and thresholds, risks and so on as mentioned in Chapter 1 should also be taken into account if a selective science policy of fostering particular fields and types of products is adopted.

Incidentally, if new products and inventions are one of the main ways in which a nation can benefit from technological progress, it could be that this implies an optimum rate of growth in the economy. At a low rate of growth of an economy R & D effort may be restricted and little innovation may occur. These are likely to rise with a higher growth rate. On the other hand, attempts to enforce very fast growth rates may result in reduced gains because more marginal advances are adopted and existing management may become overextended and taxed by trying to promote too wide a range of new products and loose efficiency.[21] Up to a point, scientific effort for industry is likely to be determined by an economy's growth rate.[22] The relationship might be broadly like that shown in Fig. 2.3.

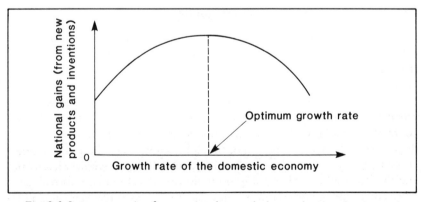

Fig. 2.3 Important gains from national growth due to the introduction and discovery of new techniques.

2.4.5 *Competition policy and science*

Although at first glance competition policy for industry may appear to be remote from science policy, they are interconnected in several ways. The degree and nature of competition in an industry can influence its level of R & D, its rate of innovation, and the rate of diffusion of new techniques. The level of benefits obtained from a new technique as well as the distribution of gains in addition may depend on the state of competition in an industry. The structure of the industry in terms of the size of firms in it, their market shares and the inequality of these shares can also exert an influence upon these factors. Economists are systematically exploring these inter-relationships and there is a considerable volume of literature dealing with these matters.[23]

Optimal competitive policy for maximizing national gains from domestic innovations and inventions can require a different strategy in the domestic market to that adopted in relation to exports and imports. This is illustrated by past Japanese policy of reducing competition between domestic exporters of technology to obtain sales and reducing competition between potential Japanese purchasers of foreign technology. It is also illustrated by the way in which some innovations have been made in Japan. Typically, on the introduction of an innovation into the Japanese economy, competing imports have been restricted and use of the innovation or invention has been restricted to one Japanese firm until its sales have exhausted economies of scale. Once economies of scale have been exhausted by the original innovator, the license to use the innovation is extended to another company and so on, thereby increasing competition in the domestic market.[24] Competition appears in such cases to be *phased competition* directed to perceived national advantage.

2.5 Import of science *vs* its local supply

It is often argued that it is more profitable to import scientific knowledge, especially basic knowledge, than to produce it at home.[25] It is suggested that from the selfish point of view of a country's own economic gain, it pays a country to 'free-ride' on the scientific effort of other countries. This, to me, seems a narrow and in some respects unrealistic point of view. New scientific knowledge from abroad is not instantly available to the home country because (a) some of this knowledge may be kept secret by its discoverers (at least for a period of time), (b) knowledge about discoveries abroad is likely to depend on the contacts which the home country has made and the effort which it has put into monitoring developments abroad, and (c) scientists in the home country may take time to assimilate and assemble the knowledge from abroad, the extent of the delay varying with the experience and knowledge which they have in the field and the extent to which they have been involved in the discovery process overseas. All countries depend to some extent on knowledge discovered overseas but it is unlikely to be in the interest of any to depend solely on overseas efforts. However, it is important to consider the extent to which a country should rely on imported science in comparison to indigenous or 'home-grown' science.

2.5.1 *Lags in the import of knowledge*

This question might be best approached by considering Fig. 2.4 which indicates some of the likely lags involved in the import of new technology and stemming technology. Once basic ideas or principles in an area are discovered overseas there is likely to be a recognition and transmission lag before they are appreciated in the domestic country. This is likely to depend to some extent on scientific effort at home, the domestic 'set' for the discovery in question and the degree of publicity given to the discovery abroad. The longer the recognition and transmission lag the more likely that the first technology using the new ideas is developed abroad and commercially introduced abroad. The development of technology using the ideas in the home country is likely to lag behind that overseas. Lag in the appreciation of the ideas of others is likely to lead to lags in technological development by a following country, such as say Australia. Being a 'late-starter' in the technology is unlikely to be very profitable. The late-starter will have to compete against imports of similar technology, exports are also likely to be hampered by the presence of competing technology and more importantly if any significant technological development is made it is likely to be used by overseas companies established in the field using reverse-engineering.[26] Furthermore, overseas firms established in the industry may already be protected by entry barriers built up during their earlier start. These include customer contacts and goodwill, preferences reinforced by advertising over a long period, economies of scale in production and distribution, and patent protection.

The length of the lags shown in Fig. 2.4 have been arbitrarily chosen. It is, however, important to note that the lengths of the lags can be reduced by increased effort, which of course uses extra resources. By effort, it may sometimes be possible and worthwhile to reduce the domestic recognition and transmission lag and development so that domestic technology arrives on the scene prior to the foreign technology or at much the same time. Where this is likely to be profitable it is likely to require frequent contacts overseas by scientists from the domestic country and a research input into the area by the home country.

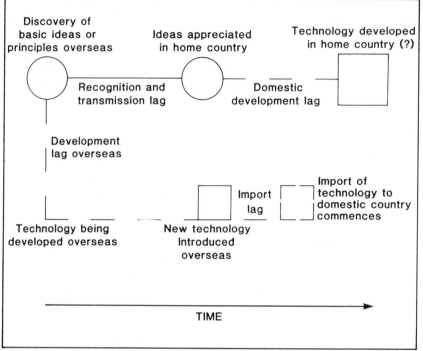

Fig. 2.4 Lags involved in the import of new knowledge and its stemming technology.

2.5.2 Import of knowledge versus its home production

Scientific and other resources are required to import scientific knowledge. For any given part of resources allocated to science, this means that the fewer resources there are for developing 'independent' domestic science and ideas, the greater is the proportion of resources devoted to the import of science. This suggests that there may be an optimal allocation of scientific resources between those to be used to foster the import of science and those to produce ideas at home, and that for any allocation of resources to science there is a trade-off between the quantity of ideas imported and the quantity produced at home.[27] The trade-off relationship might be like that indicated by curve ABCD in Fig. 2.5.

The shape of this curve indicates that the production of some ideas at home (allocation of some funds to domestic R & D) is likely to be conducive up to a point to the import of a greater number of ideas from abroad.[28] In the case shown it is possible to move from point A to point B by allocating more resources to domestic research thereby raising the quantity of ideas imported and the quantity produced locally. To move more resources into local R & D is to reduce the number of ideas imported from

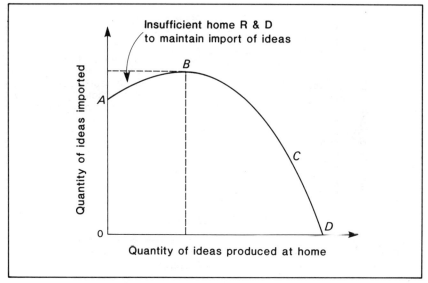

Fig. 2.5 Illustration of the possible trade-off between the import of ideas and their local production. (First approximation).

overseas although the number of ideas produced at home continues to rise.

However, it seems to me that it can be convincingly argued that curve *ABCD*, the import-domestic production science frontier in Fig. 2.5, should be modified to take account of the fact that the domestic production of ideas may be adversely affected if the import of ideas falls to low levels. This is allowed for in Fig. 2.6 by allowing the frontier to re-enter on its lower portion. The portion *CE* of this frontier indicates that as the country allocates more scientific resources towards the production of domestic ideas (becomes increasingly insular) and moves from *C* towards *E* on the frontier, the quantity of domestically produced ideas falls because there is insufficient import of new ideas.[29]

The relevant relationship between the import of science and its home production is likely to be of the type shown in Fig. 2.5. If this is so, only the allocation of scientific effort between import and domestic production of science which places the country on its frontier between *B* and *C* can be efficient. If a country is not on the *efficiency set*, *BC* of its frontier, it can by reorganizing its use of scientific resources obtain a greater quantity of one set of ideas without obtaining fewer of the other set, or obtain a greater quantity of ideas from both sources.[30] Science policy requires the *efficient set* to be at least roughly identified in practice and for policy decisions to be implemented to direct the economy to the set if it is not achieving it.

Once the efficient set for the import and domestic production of science is identified, it is necessary to determine the best combination in this set. The

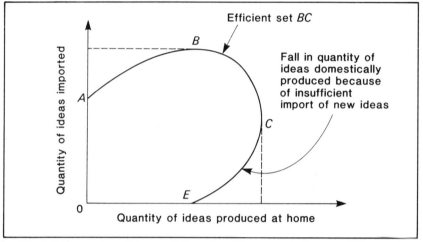

Fig. 2.6 Illustration of the possible trade-off between the import of ideas and their local production. (Second approximation).

optimal combination will depend upon the relative value of imported ideas and ideas produced at home. In the simplest case the value of imported and home-produced ideas might be represented by a series of straight-line equal value (iso-value) curves. Three of these, marked V_1V_1, V_2V_2 and V_3V_3, are shown for example in Fig. 2.7. V_1V_1 represents all the combinations of imported and domestically produced ideas which are of equal value.[31] V_2V_2 represents another set which yield an equal but higher value than the

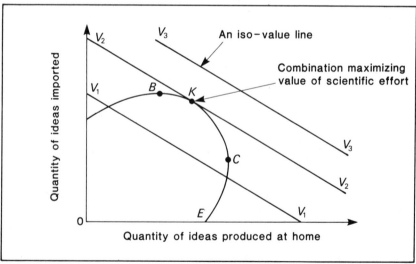

Fig. 2.7 The combination of imported and home-produced science of maximum value (K).

combination on V_1V_1. Higher iso-value curves correspond to combinations yielding a higher value. The optimal attainable combination is the one on the import/home production frontier lying on the highest attainable iso-value line. In the case shown in Fig. 2.7 it is the combination of import and home production corresponding to point K.

The slope of the iso-value curves represents the relative value of imported and home-produced ideas. The less steep these curves are, the lower is the relative value of home-produced ideas, and vice versa. The relatively more valuable imported ideas are, the closer does the optimum combination, K, approach to B. But it might be observed that iso-value curves need not be straight lines in practice. However this does not alter the substance of the argument.

2.5.3 *Selectivity in the import of knowledge*

Choices have also to be made about the areas, if any, to concentrate on in importing knowledge. Given that a limited amount of resources is available to foster the import of knowledge, it is necessary to determine the best way to use those resources. Should the resources be spread across all fields or should they be concentrated in some? What fields should be concentrated on, if any?

The effect of concentrating resources in a field is to reduce the length of the lag between the discovery of the ideas overseas and their appreciation in the home country. But the cost of concentrating resources in selected fields is fewer resources for the other fields and longer transmission lags for ideas in those fields. Unfortunately, trade-offs are involved. Broadly these might be like those indicated in Fig. 2.8.

A choice must be made if a country is to concentrate on particular fields. The value of a faster rate of transmission of knowledge in selected fields must be weighed against the gains foregone by a slower rate of transmission in other fields. While it is impossible *a priori* to specify fields in which a country should concentrate its import of knowledge, in some instances there are likely to be fields in which R & D is concentrated in the domestic economy and areas in which a country is determined to have a world lead or near world lead, for economic or other reasons.

However, it should be noted that in some fields, a country cannot rely upon the import of foreign knowledge to any great extent to solve its problems. This is likely to be so for problems which are unique to the country in question.

The import and transmission of knowledge from abroad is a most important aspect of science policy. Most countries must rely on substantial imports of knowledge, and in the case of small countries import may be by a wide margin the main source of their scientific knowledge. Autonomy or

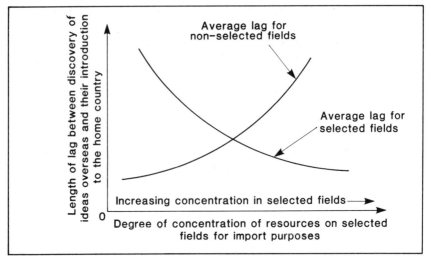

Fig. 2.8 Relationship between degree of concentration of resources in selected scientific fields and lags in the transmission of knowledge in those fields from abroad. The lags can be reduced, but this is at the expense of longer lags in other fields.

self-sufficiency in scientific knowledge is an unrealistic and uneconomic goal for a country so science policy questions concerned with the import of knowledge are likely to be of continuing relevance.

2.6 Science and social policy

While science for the purpose of supporting industry makes an important contribution to human welfare and economic development, it is worthwhile distinguishing scientific effort for (other) social purposes. It has become standard practice to separate out for special attention science for improvements in the environment, health and welfare generally.

2.6.1 *Externalities, the environment, pollution and priorities*

One of the reasons why this separation has been found necessary is that market failure tends to be widespread in activities affecting the environment, health and welfare generally. Private gains and costs associated with activities in these areas in a market system do not adequately reflect social gains and social costs.[32] For instance, motorists in a crowded city may add to the likelihood of smog and lung complaints yet not be charged (or charged enough) for the amount of carbon monoxide and other pollutants emitted by their vehicles, considering the overall damage caused;

or, to give another example, factories may emit pollutants into streams, and the pollutants may kill fish or cause other damage. Assuming that factory owners are not charged for these emissions, their private costs do not reflect the social cost of their activities. Where an individual or company is not required to reduce pollution or pay an amount equivalent to its social cost, he (it) has no incentive to purchase pollution reducing techniques unless these coincidentally also reduce his (its) private costs of operation. Consequently, it is not profitable to undertake research directed specifically at reducing pollution.

In cases where legislation is introduced to make the polluter pay (or in other ways provides legal sanctions for the adoption of less polluting techniques) for his externality costs, this is likely to give a boost to science in the area of pollution reduction because private firms now find that they have a market for pollution-reducing equipment. The increased concern of governments with pollution in the last decade and the adoption of policy measures to make the polluter pay or otherwise guide his choice of techniques has increased the profitability of producing less polluting equipment and has given a fillip to research in this area. It is not always practical, however, to charge the polluter the full cost of his unfavourable overspills. For one thing, they cannot always be easily measured. This means that government is likely to have a considerable continuing amount of responsibility for research into factors affecting the environment.

As in other fields, the government has responsibilities to support basic and other R & D in the environmental areas because of difficulties which private concerns have in appropriating gains from these. In addition, further factors may need to be taken into account. In cases where pollution cannot be practically charged for at its source, the government may support research intended to develop techniques which *simultaneously* reduce private costs and pollution overspills. Because private costs are lowered, the new techniques are adopted and society gets a bonus in the form of smaller unfavourable overspills. This is a 'second best' policy.[33]

Incidentally, increasing controls on pollution throughout the world have made the development of pollution-reduction equipment more profitable. It can sometimes be to the economic advantage of a country if it can detect or predict such social changes early and develop technology likely to be demanded due to the changes. A country may obtain an early start on research and development for such technology and, by developing a lead, become an exporter and gain in accordance with product-cycle principles.

2.6.2 *Health*

Science for the purpose of maintaining the health of the population, it is often suggested, ought to have a high priority. Markets (the price mechanism) are unlikely to encourage enough R & D in the health field because:

(a) Overspills (or externalities) are important. For instance, an individual in ill health poses a strain or burden upon those around him, is likely to be less than fully efficient in his job and could pose a risk to others, e.g. if driving a motor vehicle and he collapses.
(b) Basic techniques and principles relating to health and medicine cannot be appropriated by their discoverers and can easily be copied without payment.
(c) All run a risk of ill-health and of suffering from particular ailments. This means that most desire to support research in these areas *in case* they contact the medical ailment or problem and are the beneficiaries. The desire to do so is likely to vary with the likelihood that one will catch the disease or be a victim and the seriousness of the disease. Research in these areas is in fact a collective 'insurance' but there is no market for it and, as in the option demand case,[34] market failure occurs. Less money is likely to be voluntarily subscribed for collective research in the health area than is collectively desired and thus there is a role for the government to support research in this area.
(d) The benefits of advances in this area are likely to flow to many generations.
(e) It might also be argued that income, reflected in the prices which individuals can pay for the application of medical research, should not be the sole arbiter of desirable medical research.

If, however, the government is to fund medical and health research on a large scale it does need to establish priorities between competing claims for medical and health research. This is an area in which there is need for discussion and development of the basic principles of allocation even though this cannot be attempted in this general survey.

Once again, it may be necessary for a country to pay some attention (but not an overriding one) to the techniques which can flow from medical and health research and which are marketable. Various types of 'hardware' (mechanical and electronic aids, etc.) and drugs may be marketable and in some instances can be used to advantage to boost exports if an optimal lead is obtained.

2.6.3 *The disadvantaged*

Many of the comments made in relation to health also apply to research to assist the disadvantaged. In their case, however, it would seem particularly inappropriate to apply the ability-to-pay principle as a yardstick of the amount of desirable R & D intended to assist them. Once again this is an area in which basic research is needed on priority assessment.

2.6.4 *Energy and other issues*

It is questionable whether research on energy matters should be treated under the heading of science and social policy, but the use of energy is fundamental to our present social and industrial system. Given the depletion of non-renewable fuels, it is clear that research into the saving of energy and new sources of energy must remain a high priority. Government support for such research is required to some extent because those undertaking the research are unable to appropriate an adequate share of the gains to give sufficient encouragement to such research. Furthermore, other grounds for assistance may exist such as uncertainty of returns and large capital requirements which are difficult to meet.

Research into energy sources may also be a matter of concern for a country from the point of view of its defence, if, like South Africa, it is without oil reserves. Apart from the defence issue, most countries desire some energy reserves or alternative energy sources to those imported to cushion them against possible instability in the availability of supplies from abroad.[35]

In science policy, it is also necessary to take into account in setting priorities, other social factors such as working conditions, employment and public concern about particular scientific developments which put the general public at risk. Science policy may be designed to help improve these conditions.

2.7 **Research for defence and big science**

Defence has been described by some economists as a 'regrettable necessity'[36], but even in the case of regrettable necessities, economizing is possible. Where a specified result or invention is required, it is possible to aim to achieve this at minimum cost within a specified period of time. Alternatively, if a given quantity of funds is available, it may be possible if output or results can be indexed, to aim at maximizing output or results for the given expenditure.[37]

In planning R & D expenditure on defence and big science, a government needs to take account of favourable side-effects which can be achieved. Performers of R & D in these fields may obtain valuable 'spin-off' in the form of ideas for commercial products. While this should not be an overriding consideration in allocating research for defence purposes, it is a worthwhile consideration and could, for instance, be a factor in deciding whether to contract the research out to private industry, or to modify the type of research, or to undertake the research at home or contract it out to an ally abroad. Research undertaken at home is likely to give spin-off to local firms but not all countries have the capacity to conduct all of their research for defence purposes. In such cases they are likely to be heavily dependent on imported defence technology incorporated in foreign-produced defence equipment.

Apart from the possibility of new commercial products and techniques stemming from such R & D, the supply of government orders based upon it can provide local firms with valuable experience in new fields. Such orders provide some protection to local firms and may enable them to gain experience in production fields in which they have no previous expertise.[38] The period during which firms are meeting government contracts gives them time for learning and this may enable them to reduce their production costs and eventually to be in a position to compete successfully against overseas firms.

To some extent big science (for instance, R & D for nuclear energy) is supported by governments for purposes of national prestige and to satisfy curiosity. Various market failures can also provide a case in favour of a government supporting big science.[39] Factors such as the inability of firms to raise the large amount of capital required, the uncertainty of results, the fact that much basic research is involved and benefits from results are only imperfectly appropriated, can be used to support the involvement of government in big science with possible civilian applications. Possible gains do have to be carefully assessed because large losses can be made by a country if it attempts to develop big science projects, the products of which are subsequently not in demand. The development of civilian supersonic transport by the UK and France appears to have been a case in which considerable national losses have occurred because demand for such transport is lower than predicted.

2.8 Basic *vs* applied *vs* developmental science

As mentioned earlier there does not appear to be any *a priori* way of determining the ideal balance in a country between basic, applied and developmental science although the trite observation can be made that science does not add to material wealth and economic welfare until it is pursued through to the development stage and results in new products and techniques which are actually used. Nevertheless, it is possible to consider some of the factors to be taken into account in searching for an ideal balance.

There is likely to be considerable complement between activity in basic and applied research and the development of new products and techniques. Development and applied research is dependent to a large extent on the stock of ideas stemming from basic research. As C.P. Snow has suggested, a country which seriously depletes or fails to add to its stock of basic science may also soon find that its supply of new products and techniques falter. From an economics point of view, there may well be bounds upon the ratios of basic, applied and development research likely to be optimal in terms of the quantity of useful inventions produced.

These bounds can be determined in principle by drawing upon economic theories of production.[40] In a particular economy the bounds might be like those indicated in Fig. 2.9. Only combinations of expenditure on basic and applied/development research in the hatched area are efficient in terms of the quantity of inventions produced. Combinations outside this region involve greater expenditure on science or research than is needed for the results achieved.[41] Combinations in region A involve too much expenditure on basic research relative to applied in the sense that the *same output* in terms of quantity of inventions produced could be obtained with less expenditure on basic research and no greater expenditure on other types of research. Similarly combinations in region C involve relatively too much expenditure on applied research/development.

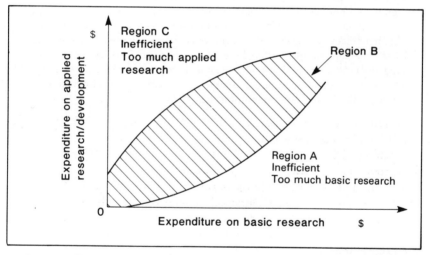

Fig 2.9 Efficient combinations of expenditure on basic research and applied research/development (hatched area) in terms of the quantum of inventions.

Fig. 2.9 allows for the possibility that there can be too little development or too little basic research to maximize the output of inventions and ideas for a given expenditure on science.

The above discussion is, however, based on the so-called 'overheads' doctrine of basic research. This is the view that the case for government support for pure science rests upon 'the contribution which it makes, as an overhead or investment towards other national goals, such as defence, productivity and health.'[42] Toulmin, however, believes that this approach poses a problem, a problem which he states as follows:

> The essence of the resulting problem is this: if the fundamental argument for government support of basic research remains entirely the prospect of economic payoff – even *indirect* economic payoff – then decisions about the subdivision of the total available research funds between different lines of basic research will cease to be a matter for pure scientific judgement alone and become an economic, or even a political, issue.[43]

Another point of view expressed in favour of basic science is that it is an expression of 'high civilization' and should be supported as a cultural activity or merit want. Possibly some of the most cutting comments against this view have been made by economist, Harry Johnson. He has said:

> The concept of 'scientific culture' raises a number of questions, among which the most fundamental is the question whether basic scientific research is – in the economist's terms – to be regarded primarily as a consumption or an investment activity . . . Much of the contemporary 'scientific culture' argument for government support of basic scientific research is such as to put it – intentionally or not – in the class of *economically functionless activity*. The argument that individuals with a talent for such research should be supported by society, for example, differs little from arguments formerly advanced in support of the rights of the owners of landed property to a leisured existence, and is accompanied by a similar assumption of superior social worth of the privileged individuals over common men.[44]

Toulmin's defence of basic research is rather disappointing and is less satisfactory in my view than the overheads doctrine. He defends basic research on the basis that it leads to interesting *employment*.[45] He accepts the overproduction doctrine that man's production of and possibilities for producing material goods are out-stripping his possibility of using them and if employment is to be maintained, an increasing number of individuals must be employed in 'unproductive' pursuits such as basic research and cultural activities. His view cannot, however, be dismissed out of hand.

As research develops from the basic stage, to the applied stage and to the development stage the results of that expenditure become more fixed or inflexible. Basic research is flexible and not to any great extent *embodied* in equipment but development work is as a rule embodied in equipment suitable only for a narrow range of tasks. An implication of this appears to be that more and more economic and other knowledge is needed, if economic losses are to be avoided, as the development stage is approached, especially since development work is likely to be the most expensive component of a research project. This is illustrated in Fig. 2.10. Clearly development work should be carried out by the economic agents who have

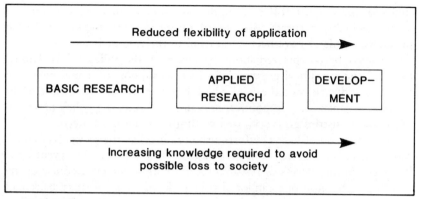

Fig 2.10 The relationship between flexibility and risk and the sequence from basic research to development.

the most knowledge in the economy about its economic and other prospects. Some individuals argue strongly that this is private industry whereas others contend that, at least in some instances, the government has superior knowledge. However, it is pertinent to note that if governments undertake the lion's share of development, they may be taking 'bread and butter' (profits) away from private industry. Above-normal profits are the usual reward to industry for successful development and innovation. Society is faced by a dilemma. While development involves greater risks for society than basic research, basic research must be followed through to development and application if it is to have economic value.

In some cases, there is no alternative to development work being undertaken or financed by the government. This is the case for instance when new products or processes cannot be protected by patents or cannot be effectively protected by them. For example, new crops and improved breeds of farm animals and some agricultural techniques cannot be protected by patents and there is a case for governments to undertake development work in these areas.

2.9 Concentration and dispersion of R & D effort

2.9.1 *Desirability of concentration on selected fields*

The desirability of concentrating on selected fields of science and science for selected industries or products has already been discussed to some extent in considering the product-cycle hypothesis and the import of science.

Governments in Japan and more recently in Britain have adopted policies of favouring particular industries and classes of projects in allocating their funds for the support of R & D. In the case of Japan this policy, combined with reasonably accurate forecasts of future demand for new types of products, seems to have been successful. In the case of Britain, however, doubts have been expressed about the success of the policy.[46] It is claimed that in Britain political pressures have led to support of 'lame-duck' and declining industries and these have had their R & D heavily subsidized. In contrast in Japan, despite political pressures at home, the policy of letting declining industries go (very often offshore) has been followed.

Concentration of R & D efforts although it offers the prospects of substantial gains also brings considerable risk. It is impossible to predict the likely economic success of research efforts with accuracy. Because of the risk factor, because one cannot always pick winners and may back bad losers, a case can be made out for some government support of R & D effort generally (across a wide spectrum) with some extra support being made available to selected industries and priority projects. This policy means that not 'all eggs are placed in the same basket' and yet some allowance is made for product-cycle and product/technique leadership considerations.

Of course R & D efforts in industry do tend to be naturally concentrated. They tend to be concentrated in so-called science-based industries such as chemicals, aviation, engineering, electronics and transport equipment. The above discussion is, however, concerned with concentration that is deliberately fostered by the government.

2.9.2 Duplication of R & D effort

While it is sometimes suggested that duplication of R & D effort is wasteful, it need not be wasteful especially if desired results in the field have a high priority. As a rule the net effect of duplication, where each research effort is funded at the same level as otherwise, is to increase the probability of an *earlier* success from the research. This is because:

(a) Competition between the different research groups may spur each group on.
(b) Each research group may adopt a different method or approach to its research; (*a priori* it may not be clear which method will succeed and how quickly any of the methods will give a breakthrough).
(c) Different research approaches may provide different techniques to choose from and provide different side-results and inventions of value.

By combining the results of Scherer and Mansfield,[47] it is possible to consider more formally the question of time saved by duplicating R & D effort by using a figure like Fig. 2.11. In Fig. 2.11 curve AB indicates the amount of time required for a breakthrough in research as a function of the amount of funds provided for that research when the research is undertaken by 'one team'. The dashed curve *CD* indicates the time required for a breakthrough when the research funds are equally divided between two independent teams. The figure indicates that at low levels of funding for the research, duplication increases the time required for a breakthrough in the research but at higher levels of funding duplication reduces anticipated breakthrough time. Breakthrough time when funding is low is likely to be greater for duplication than for non-duplication because of the existence of resource thresholds and some economies of scale in research projects. For a project adequately funded in which a quick breakthrough has a high priority, duplication of research effort is likely to be optimal. In the case illustrated in Fig. 2.11, duplication of research effort is the better strategy when available funds for the overall effort exceed *OM* and time is the major consideration.

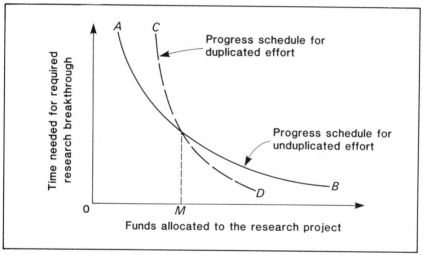

Fig. 2.11 The efficiency and inefficiency of R & D duplication effort.

2.9.3 *New fields*

Special difficulties are involved in ensuring that sufficient R & D is undertaken to support new and developing scientific fields and industries. Research expenditure, especially in industry, is heavily influenced by *existing* economic patterns. Firms generally undertake research in areas (or areas close to) of their existing production expertise and their funding for these is normally a relatively constant percentage of their sales receipts. Furthermore, political pressures for the allocation of government research grants are likely to be most intense from established industries and institutions and researchers in established fields. Therefore, governments

62 Science and Technology Policy

need to give special consideration to the establishment of mechanisms to encourage (and 'vet') new fields of science possessing economic potential.

2.9.4 Regional distribution R & D

The geographical distribution of science effort within a country is important. In this respect Professor Harry Johnson has said:

> It seems desirable to draw attention to a facet of policy toward basic science that is important but tends to be overlooked by scientists. This is the implication of the geographical distribution of science support for the pattern of growth of an economy. The location of scientific research activity in a particular city or region generally constitutes a focal point for the development of science-intensive industries in the surrounding area. This should be taken into account in deciding on the location of such scientific activity. There is a natural tendency for scientific activity to agglomerate around established centres of scientific accomplishment. This is probably the most efficient way of conducting scientific research from the point of view of science itself. From the economic and social point of view, however, and perhaps even from the longer run scientific point of view, there is a strong case for encouraging the development of scientific research centres in the more depressed and lower income sections of the country, as a means of raising the economic and social level of the population in those sections. Much of the poverty problem is associated with geographical concentration of high-income industries in certain areas and their absence from others, which makes migration the only feasible route to economic improvement. A deliberate policy of locating scientific research in the backward areas of the country to encourage their industrial development could in the long run provide a socially and economically more attractive attack on the poverty problem than many of the policies now being applied or considered.[48]

There are some types of research which can only be effectively carried out in regions where the research is expected to be of benefit. For instance, research into tropical agriculture in Australia (or much of this research) can only be effectively carried out in tropical regions of Australia. Research into some industrial and environmental problems may need to be carried on in the locality where the problem occurs.

When considering the possible location of a science or research activity, consideration needs to be given to the local 'spin off' which it may engender. It may stimulate local industry and have a local income-multiplier effect.[49] Decentralization of research activity can play a role in helping to decentralize industry and population but in some instances this policy can be costly because of the loss of economies of agglomeration which can be obtained in larger cities.

It is frequently suggested that useful feedback and economies can be achieved by having universities and research bodies, both government and private, all located in the neighbourhood of one another. These 'science cities' need not be located in an existing city but could be located some distance from it.[50]

Another method of maintaining some concentration yet spreading research around a country is to establish different 'research centres of excellence' in different regions. Even here there can be arguments in favour of duplication of research areas of excellence and some attention needs to be paid to outstanding individual scientists who are not prepared to move for extra funds to the region designated as outstanding in their speciality.

2.10 **Performers of R & D**

The social benefit obtained from research may vary (depending on the type of research) with the group performing it. This benefit for instance, may vary depending upon whether the research is performed by the government, universities or private industry and account needs to be taken of this when allocating research funds.

2.10.1 *Government vs universities vs private industry as performers*

There seems no good reason to concentrate the performance of R & D in either government, the universities or industry. Each has a role to play in relation to different types of research. However, it would be theoretically possible for instance to concentrate all R & D in industry by the government contracting out its research to industry.

Arguments against such a policy include the following:

(a) In the case of basic research, it is difficult to check whether the contract is being fulfilled.
(b) Research bodies in industry may not have the long-term permanency of those in government or the universities and therefore there is a greater risk that expertise in an area will be lost.
(c) In general, industry does not have the overheads (libraries, equipment, etc.) available to universities and government.
(d) Research contracts from government may be only used by firms in industry to 'top up' their research programmes as required, that is when they develop excess capacity in their research divisions. Thus government planned research may proceed by 'fits and starts'.
(e) Those undertaking government-sponsored research may obtain spin-off in the form of patentable inventions. It might be argued that they are given an unfair advantage unless payment is made to the government for these side-benefits.

Arguments in favour of such a policy of contracting out research to industry include the following:

(a) Industry obtains valuable additional experience in research which can be used on other occasions to its own benefit.
(b) Firms may obtain more spin-off than would be the case if the research is conducted within government. They are more likely to perceive commercial uses and discover side-inventions of commercial value.
(c) This policy enables the research facilities of industry to be more fully and evenly utilized.
(d) In general the more research undertaken by industry the more likely is research to be of commercial value, because industry is the best judge of the commercial viability of a project.
(e) This policy enables a wider range of expertise to be drawn on than would otherwise be available to a single organization.

2.10.2 Co-operative research

It is sometimes suggested that it is better for industries to undertake their own co-operative research rather than entrust this to governmental institutions. To this end, industries sometimes set up central research organizations funded by contributions from firms in the industry.

Unless the research is purely on a contract basis (and in this case, it is not co-operative research) there are likely to be problems in funding collective research. If contributions from firms in the industry are voluntary some firms may 'free-ride', some firms may not contribute or contribute their fair proportion to co-operative research and may still be able to get access to research results of a general nature from the co-operative research institution. United Kingdom evidence[51] indicates that in the case of voluntary research associations, the main financial contributions tend to come from the largest firms in an industry and that these firms tend to dominate the activities of the research association. Most research may be along the lines or in general areas suggested by large firms and suggested with an eye to the possibility of their capitalizing on such work.

Work by Johnson[52] in the UK indicates that such research associations rarely patent inventions. Their work generally stops short of a patentable invention and one or more of the contributing firms to the co-operative carries any useful ideas through to the stage when they result in a patentable invention. Clearly few firms in secondary industry are likely to agree to collective central research on their behalf if it results in patentable inventions made widely available by the central research organization. The profitability of invention and to a large extent above normal profits for firms in industry depend upon individual firms obtaining inventions before others and thus gaining a temporary monopoly.

2.10.3 *Domicile of performers of R & D – foreign-owned firms*

In some countries, for example Canada, preference is given to domestically owned firms when distributing government R & D contracts and subsidies for research. This is done in the belief that it is likely to mean greater gains for the country sponsoring the research. It is argued that preference for domestically owned firms ensures that (a) any extra profit resulting from the R & D adds to domestic income and is not transferred to shareholders abroad; (b) the technology resulting from the R & D is likely to be first used at home and this gives product-cycle advantages to the domestic country and (c) the expertise of local firms and their barriers to foreign firms are raised and this is important in the cumulative competitive process whereby technological differences between firms tend to be magnified by time. Any technological advantage gained by foreign-owned firms (for instance as a result of subsidization of their research effort in a host country) and not available to domestically owned firms, tends to increase the technological gap between them with time.

On the other hand, it is possible that:

(a) No locally owned firm is capable of undertaking the R & D required or can only do so at exhorbitant cost.
(b) Foreign-owned firms may in fact obtain much greater sales for the technology through overseas subsidiaries and the domestic economy may thereby gain by higher wages and tax receipts if the new technology is produced domestically.
(c) Foreign companies may not in fact repatriate profit but may continue to plough most of the profit of their subsidiaries back into investment in the host country.
(d) Support of R & D by foreign-owned firms can be regarded as a subsidy helping to encourage capital inflows.
(f) Research by foreign-owned firms may result in beneficial overspills for local firms. For instance, it may result in more highly skilled researchers who might alter their employment to domestic firms at a later stage.

The issues involved are extremely complex and discussion is complicated in some instances by emotionalism. More research is needed in this area. However, even given the present state of knowledge it is clear that subsidization of the R & D activities of foreign-owned firms and their equal access to government R & D contracts can pose a dilemma from the point of view of maximizing national gain. Ideally, each case needs to be considered on its merits. However the problem with this is that it can involve too much 'red tape' and on balance be detrimental to the national interest. More study of basic relationships is needed before general guidelines are advanced for favouring domestically owned producers over foreign-owned producers in R & D grants and contracts.

2.11 Service science

I regard service science as covering scientific activities or resources which service or provide inputs into the R & D activities of a range of researchers and scientists or provide diagnostic-type services. Most service science is provided by government and universities. Examples of service science are library services, the availability of information about patents and the widespread availability of sophisticated testing and analysing services or equipment to the public or a segment of it.

The arguments in favour of service science being provided by governments are the traditional ones put forward to favour public enterprises (over private ones) in special circumstances. It is argued that public provision is likely to be most efficient if:

(a) Duplication of facilities would be wasteful and duplication might occur under free enterprise.
(b) A private monopoly would otherwise occur or is likely to occur because of natural factors such as decreasing per unit costs with increasing volumes of business.
(c) The socially most beneficial operation of the facility involves its operation at a loss.
(d) The required subsidy to ensure the operation of the service facility by private enterprise is so high that problems are involved in public accountability for the public funds involved in the subsidy.

It is not possible to consider the operation of these principles in relation to all types of service science here. Take the case of libraries, however. These would seem to qualify for government provision on a number of bases: (a) decreasing per unit costs occur with increased size as measured by volume of borrowings; (b) borrowings produce favourable overspills to the rest of society and a considerable subsidy could therefore be justified. Also consider the case of expensive testing or analysing equipment. Economies of scale in its use may be such that only one piece of equipment of this type is required in a small economy, such as the Australian, if costs per test performed are to be minimized. A private owner could, therefore, obtain a monopoly which he might exploit up to a point and this provides an argument for public provision of the testing facility.

The arguments against government provision are also the traditional ones. For instance, it may be argued that public enterprises tend to be inefficiently managed. Nevertheless, in all countries governments provide some service science facilities or in some cases regulate their provision by private enterprise, for instance, by fixing maximum prices to be charged for particular types of tests or diagnoses using private equipment.

In passing, it might be noted that economic research into service science and public enterprise has been relatively neglected. Research in this area is in need of encouragement. As mentioned earlier, the fact should not be overlooked that the research facilities of government bodies and universities can frequently be used more efficiently when their use is shared by others. Shared use may eliminate uneconomic duplication costs, and ensure that existing facilities are operated at more economic levels. Even though there are limits to the extent to which sharing is possible, greater consideration needs to be given to this issue.

2.12 Science and international affairs

Scientific activity provides scope for co-operation between nations as well as intense competition between them. Even when nations co-operate in fundamental research, application-oriented aspects of it throw some shadow of competition on the process. As King puts it: 'International scientific activity is a blend of co-operation and competition'.

2.12.1 *International co-operation in science*

There are a number of scientific fields in which nations find it worthwhile to co-operate. These include:

(a) Fields of general interest or fields with non-marketable applications extending beyond any single country. These include fields like astronomy, oceanography and meteorology. Aspects of agricultural research of common interest to several countries such as tropical agriculture and disease control can also give scope for co-operative scientific effort.

(b) In the field of 'big science' such as civil nuclear energy research or satellite and space programmes, smaller nations are unlikely to be able to participate because of the cost involved unless a number co-operate.

(c) It is possible that in new fields of research, expertise and experts may be scattered. Nations may co-operate to ensure that these experts can meet regularly and exchange ideas.

(d) Countries may be able to avoid wasteful duplication of effort in fields in which they are only marginally competitive. As King suggests:

A number of fields of applied research, particularly within the public domain, are only marginally competitive and, as problems are common to many countries, organization on a nation-by-nation basis involves a high degree of duplication. There is a strong argument, therefore, for the establishment of programmes either internationally developed or in common to adjacent countries in a region. An example is road research.[53]

Where international bodies are specifically set up to co-ordinate the joint research projects of nations, these have to be funded and careful consideration has to be given to this in considering priorities in the allocation of available national funds for the support of science. In this regard Alexander King makes the following pertinent points:

> The essential need is perhaps for countries to regard their international research commitments more deliberately in terms of the national science policies. Contribution to international schemes are, in the end, justified if they provide an extension to domestic research resources or if they enrich these or remove a proportion of unnecessary duplication. If this is to be a dominant criterion in deciding whether a country will or will not adhere to a particular international research scheme, it will be essential for the country to possess an internal mechanism for discussion of international research co-operation and its function of complementing work within the country. Yet in extreme cases the finance for an international project may come through a government department different from, and with little contact with, the department responsible for financing domestic effort in the same field. In many instances also, there is no real co-ordination of policies towards the different international organizations.[54]

2.12.2 *Information, science and international affairs*

A country's ability to participate fully and meaningfully in international affairs may vary with its scientific knowledge. A nation for instance with little knowledge and experience in the nuclear energy field may have no voice in many international bodies dealing with nuclear energy. Decisions by such bodies could influence its affairs, especially if it happened to have uranium ore for export.

Adequate scientific and economic knowledge is likely to be essential when international commodity agreements are being negotiated, for instance uranium sales, if a country is to strike the best realistic bargain. The same is true for agreements covering the import and export of technology.

In certain cases, nations may back up their special interest claims in a territory by conducting scientific research there. Antarctica is a special case in this regard.

It is likely to be in the national interest to foster flows of scientific information from abroad. This can be done in many ways. These include visits by nationals abroad to study developments at first hand, exchange schemes, better world inter-library loan schemes, the collection of information by science attachés and so on. The discussion in Section 2.5 indicates that in small relatively dependent economies, a high priority should be given to this aspect and policies in this area should be carefully and specifically formulated.

2.12.3 Scientific aid to developing countries

Many developed countries believe that they can assist the development of less developed countries by providing them with scientific and technological aid. Emphasis has recently been placed upon this aspect by the United Nations in its programme for the *Second Development Decade*.

There are numerous ways in which developed countries can be of assistance to LDCs but assistance is not a straightforward matter. Developed countries may assist, for instance, by providing training for scientists, engineers and technical staff and most developed countries have provided some assistance in this way. Such assistance, however, may raise a balance problem. There may be little demand for scientists and engineers, especially those with sophisticated Western training, and the demand may be mostly for individuals with technical skills. It is important to determine where the real need for assistance with skills is, otherwise the type of training given may be inappropriate.

Developed countries may be able to assist with on-the-spot applied research and development designed to meet the special problems being encountered by an LDC; but in general the request for help must come from the LDC and the donor must be sure that there is genuine political support for the project.

Aid-giving raises complex political issues and varying motives. While donors may give aid to LDCs for philanthropic reasons, they may also do this to obtain political goodwill and to foster their own self-interest. Boxer[55] lists the following as possible benefits to a donor country from technical and scientific assistance to an LDC:

(a) Improvements in facilities in a host country which are subsequently used by the donor country. Improvements in air communications may be brought about by aid from a donor. The donor's international air traffic may make substantial use of the improved facilities.
(b) The assistance may develop contacts and help to open up potentially profitable markets.
(c) Certain projects may be chosen because these will result in orders for equipment from firms in the donating country.
(d) The aid may give useful experience to participants from the donating country.
(e) The aid may prove to be of strategic advantage to the donor. For instance, it might improve the defence capacity of an LDC regarded as an ally.

It should be noted that self-interest on the part of the donor does not rule out the possibility of gain to the host country from the aid. Mutual gain is possible when parties act in their own self-interests.

The best form in which to provide aid needs to be kept continually under review. It cannot be assumed that scientific aid, particularly if it is to support curiosity-type science, is the best form of aid for an LDC in which, for instance, illiteracy is still a major problem. While there is a role for scientific and technological aid, miraculous effects cannot be expected from such aid if its use is not supported and complemented by appropriate economic and social conditions.

A brief mention should once again be made of pessimistic views about the impact on LDCs of science from developed countries. Many adherents of the centre-periphery theory believe that contact of LDCs with the science of developed countries helps to perpetuate their economic backwardness and political dependence.[56] For example, if technology is transferred by the direct investment of multinational companies, it may be inappropriate to the LDC's labour/capital availability and may create an enclave solely concerned with producing goods for export to developed countries. There may be little or no benefit to the developing country as a whole.[57] Again by way of emulation, academics in LDCs may try to concentrate on the frontier glamour fields of enquiry of developed countries. These fields may give little economic returns even in developed countries. A number of neo-Marxists predict that capitalist countries using advanced science and technology exploit LDCs and that living standards in LDCs are not likely to improve, indeed may worsen.[58] This view is in conflict with the neo-classical economic prediction that increased world trade, greater exchange of ideas and international investment will raise living standards in LDCs and are likely to promote equalization of incomes throughout the world.

Looking at World Bank statistics, neither the Marxist prediction nor the neo-classical economic one seem as yet to have been fulfilled. Between 1968–1978 GNP (Gross National Product) per capita rose at an annual average growth rate of 1.6% for low-income countries (excluding capital-surplus oil exporters), 3.7% for middle-income countries (excluding capital-surplus oil exporters), 3.7% for industrialized countries, 7.1% for capital-surplus oil exporters, and 4.0% for centrally planned economies.[59] Incomes per capita in LDCs did not decline but rose and a larger population was supported. At the same time life expectancy substantially rose and illiteracy was considerably reduced in LDCs. On the other hand, the gap or difference in per capita real GNP between industrialized and less developed countries widened *on average* using these statistics, but in developing countries encouraging foreign investment and geared to foreign trade and new technology such as Taiwan, Brazil, Korea, Philippines, Singapore and Thailand, greater growth rates were achieved than in other LDCs. The rates were higher than in industrialized countries so these developing countries narrowed the gap in real income. Hence, adherents to the neo-classical view may believe that this helps support their thesis. Yet one may challenge the adequacy of the quoted World Bank statistics as a basis for international comparisons of living standards. Adjustments could alter the balance of the argument either way.[60] My main purpose here is not to try to resolve this issue (it is a complex one) but to bring attention to it.

2.13 Some concluding comments

The setting of science priorities involves a consideration of wide issues. Indeed these issues may be so wide-ranging that they cannot be fully appreciated *in toto* and the policy-maker may to some extent be driven back to the piecemeal type of approach recommended by Lindblom[61] and discussed in Chapter 1. Nevertheless some progress has been made in understanding science policy issues. Greater knowledge about science policy options can be advantageous even in a piecemeal policy process. Better understanding of issues is likely to improve the quality of the input into the decision-making process by the various parties involved in policy-making.

The above discussion has also highlighted the need for further research into priorities and science policy. Some of the suggested areas for further research into priorities and science policy are:

(a) in service science;
(b) in medical science and science for the disadvantaged;
(c) science and national gains in terms of the product-cycle and technological gap hypotheses;
(d) the preferential treatment of domestically owned firms in comparison to foreign-owned firms for R & D contracts and grants;
(e) the optimal concentration of scientific efforts;
(f) the best use of resources for the import of science.

There is also need for continuing work in other areas such as the regional impact of science, and choices to be made in international co-operation in science and the role that science can play in the development of LDCs.

While advances in science provide a possible means for economic and social development, their sole benefit may be to satisfy our curiosity unless they are embodied in new technology directed to economic and social ends. The next chapter investigates the links between science, technology and economic change and the ways in which governments may intervene to promote or alter technological change, a process commonly believed to stem from scientific advance.

Notes and references

1. Carter, C. F. and Williams, B. R. (1964), Government scientific policy and the growth of the British economy, *The Manchester School of Economic and Social Studies,* **32**, p.198.
2. As far as I am aware there have been no empirical studies of the extent to which

different educational institutions in fact lag or lead in the use of techniques.
3. See, for instance, Iinuma, J. (1973), The introduction of American and European agricultural science into Japan in the Meiji era, in: *Technical Change in Asian Agriculture* (ed. R. T. Shand), Australian National University Press, Canberra, pp. 1–8.
4. Oshima, K. (1973), Research and development and economic growth in Japan, in: *Science and Technology in Economic Growth* (ed. B. R. Williams), Macmillan, London, p. 318.
5. Oshima, K. (see ref. 4) p. 323.
6. Carter, C. F. and Williams, B. R. (see ref. 1) p. 197.
7. See, for instance, Griliches, Z. (1958), Research costs and social returns: hybrid corn and related innovations, *The Journal of Political Economy*, **66**, pp. 419–31.
8. Ben-Porath, Y. (1972), Some implications of economic size and level for investment in R & D, *Economic Development and Cultural Change*, **21**, pp. 96–100.
9. Ben-Porath, Y. (see ref. 8). p. 98.
10. Carter, C. F. and Williams, B. R., (see ref. 1) p. 199.
11. Eads, G. and Nelson, R. R. (1971), Governmental support of advanced civilian technology: power reactors and supersonic transport, *Public Policy*, **19**, pp. 405–427 (extract: p. 426).
12. Oshima, K. (see ref. 4) p. 320.
13. See, for example:
(a) Kemp, M. C. (1955), Technological change, the terms of trade and welfare, *The Economic Journal*, **65**, pp. 457–69.
(b) Duncan, R. C. and Tisdell, C. A. (1971), Research and technical progress – the returns to the producers, *The Economic Record*, **47**, pp. 124–29.
(c) Bhagwati, J. (1958), Immiserizing growth: a geometrical note, *Review of Economic Studies* (June).
14. See discussion later of Japanese policy in this regard.
15. Contributors to the new technology theory of international trade include:
(a) Posner, M. V. (1961), International trade and technical change, *Oxford Economic Papers*, **13**, pp. 323–41.
(b) Freeman, C. (1963), The plastics industry: a comparative study of research and innovation, *National Institute Economic Review*, (Nov.), pp. 22–62.
(c) Hirsch, S. (1972), The United States electronics industry in international trade, text in: *The Product Life Cycle and International Trade* (ed. L. T. Wells Jr.), Harvard University, Boston, pp. 39–52.
(d) Hufbauer, G. C. (1966), *Synthetic Materials and the Theory of International Trade*, Duckworth, London.
(e) Wells, G. C. (1969), Test of a product cycle model of international trade: U.S. exports of consumer durables, *Quarterly Journal of Economics*, **83**, pp. 152–62.
(f) Gruber, W., Mehta, D. and Vernon, R. (1967), The R & D factor in international trade and international investment of United States industries, *Journal of Political Economy*, **75**, pp. 20–37.
(g) Lowinger, T. (1975), The technology factor and the export performance of U.S. manufacturing industries, *Economic Enquiry*, **13**, pp. 221–36.
(h) Teubal, M. (1975), Toward a neotechnology theory of comparative costs,

Quarterly Journal of Economics, **89,** pp. 414–31.
16. Gruber, W., Mehta, D. and Vernon, R. (see ref. 15f) pp. 20–21.
17. Vernon, R. (1966), International investment and international trade in the product cycle, *Quarterly Journal of Economics,* **80,** pp. 190–207.
18. Baldwin, W. and Childs, G. L. (1969–70), The fast second and rivalry in research and development, *Southern Economic Journal,* **36,** pp. 18–24.
19. Baldwin, W. and Childs, G. L. (see ref. 18) p. 21.
20. If a new product is rapidly imitated by others and if they have advantages such as cost or marketing advantages, the innovator may gain little or even lose while the imitators profit handsomely.
21. This is a similar theory to the one advanced by Marris in his discussion of the rate of growth of the firm and its profitability. See Marris, R. L. (1964), *The Economic Theory of 'Managerial' Capitalism,* Macmillan, London; and Wildsmith, J. R. (1973), *Managerial Theories of the Firm,* Martin Robertson, London, Ch. 7.
22. R & D expenditure may be partly a function of an economy's growth rate and with a lag may feed back to increase that growth rate. This is particularly likely for industrial R & D since many firms devote a relatively fixed percentage, e.g. 2%, of their sales revenue to R & D.
23. See, for instance:
(a) McConnell, C. R. and Peterson, W. C. (1965), Research and development: some evidence for small firms, *Southern Economic Journal,* **31,** pp. 356–64.
(b) Vernon, J. M. and Gusen, P. (1974), Technical change and firm size: The Pharmaceutical industry, *Review of Economics and Statistics,* **56,** pp. 294–302.
(c) Rosenberg, J. B. (1976), Research and market share: a reappraisal of the Schumpeter hypothesis, *The Journal of Industrial Economics,* **25,** pp. 101–12.
24. Kamien, M. I. and Schwartz, N. L. (1975), Market structure and innovation: a survey, *Journal of Economic Literature,* **8,** pp. 1–37.
25. Carter, C. F. and Williams, B. R. (see ref. 1.).
26. Baldwin, W. and Childs, G. L. (see ref. 18).
27. One of the problems with this presentation is that ideas are not homogeneous. However, the presentation is of heuristic value.
28. In other words scientific activity at home and home-produced ideas complement rather than compete with the import of ideas up to a point.
29. Up to a point the quantity of imported ideas may complement the quantity of ideas produced at home.
30. The efficient set is the set for which it is impossible to increase the quantity of imported ideas without reducing the quantity produced at home, and vice versa.
31. These curves are similar to iso-revenue curves or indifference curves in economic analysis.
32. For review of some of the relevant economic literature see:
(a) Pearce, D. W. (1976), *Environmental Economics,* Longman, London.
(b) Tisdell, C. (1978), A further review of pollution control, *Research Report or Occasional Paper* No. 44, Department of Economics, University of Newcastle, June.
33. This is 'second best' because unfavourable externalities may continue to exist and this may not be the most efficient way of reducing the overspills.
34. See, for instance, Kahn, A. E. (1966), The tyranny of small decisions: market

failures, imperfections and the limits of economies, *Kyklos*, **19**, pp. 23–47.
35. Recent Japanese science policy has been greatly influenced by this consideration. See Tisdell, C. (1975), An Australian review of Japanese science and energy policy, *The Australian Quarterly*, **47**, pp. 44–61.
36. For example, see Nordhaus, W. and Tobin, J. (1970), *Is Growth Obsolete?* Cowles Foundation Discussion Paper No. 319, December, Yale University, New Haven.
37. For a discussion of these issues, see Mansfield, E. (1968), *Defence, Science, and Public Policy*, Norton, New York, Parts 2 and 3.
38. For an interesting study of spin-off and interaction in research, see Langrish, J., Gibbons, M., Evans, W. G., and Jevons, F. R. (1972), *Wealth from Knowledge*, Macmillan, London, especially pp. 24–32.
39. These arguments are summarized in Pavitt, K. (1976), Government support for industrial research and development in France: theory and practice, *Minerva*, **14**, pp. 331–54.
40. The boundary lines are the ridge lines of production functions. See, for example:
    (a) Leftwich, R. H. (1976), *The Price System and Resource Allocation*, 6th edn, Dryden Press, Hinsdale, Ch. 8.
    (b) Tisdell, C. (1972), *Microeconomics: The Theory of Economic Allocation*, John Wiley, Sydney, p. 143.
41. In production theory such combinations correspond to the use of more resources than are needed to produce the quantity of output achieved.
42. Toulmin, S. (1965), The complexity of scientific choice II: Culture, overheads or tertiary industry? *Minerva*, **4**, p. 158.
43. Toulmin, S. (see ref. 42) p. 159.
44. Johnson, H. G. (1965), Federal support of basic research: some economic issues, in: *Basic Research and National Goals*, US Government Printing Office, Washington, p. 132.
45. Toulmin, S. (see ref. 42).
46. See Saunders, C. T. (1977), Concentration and specialization in Western industrial countries in: *Industrial Policies and Technology Transfers between East and West* (ed. C. T. Saunders), Springer Verlag, Vienna.
47. Scherer, F. M. (1965), Government research and development programs, in: *Measuring Benefits of Government Investments* (ed. R. Dorfman), Brookings Institution, Washington, and comments by E. Mansfield in the same volume.
48. Johnson, H. G. (1975), *Technology and Economic Interdependence*, Macmillan, London, pp. 26–27.
49. See, for instance, Tisdell, C. (1977), The government, education and research as factors in the development of Newcastle and the Hunter, in: *The Future Economic Prospects for Newcastle in its Region* (ed. J. Hill), Institute of Industrial Economics, University of Newcastle, pp. 123–42.
50. Japan is planning a science city away from Tokyo.
51. Johnson, P. (1973), *Co-operative Research in Industry: An Economic Study*, Martin Robertson, London.
52. Johnson, P. (see ref. 51).
53. King, A. (1974), *Science and Policy: The International Stimulus*, Oxford University Press, Oxford, p. 81.

54. King, A. (see ref. 53). p. 82.
55. Boxer, A. H. (1969), *Experts in Asia,* Australian National University Press, Canberra, p. 31 et seq.
56. See, for instance, Frank, A. G. (1971), *Capitalism and Underdevelopment in Latin America,* Pelican, Harmondsworth, and other references given in Ch. 1, note 42.
57. For a stimulating discussion of appropriate technologies for LDCs see Schumacher, E. F. (1973), *Small is Beautiful: Economics As If People Mattered,* Harper and Row, New York; and by the same author, *The Age of Plenty: A Christian View,* St. Andrew Press, Edinburgh, 1974.
58. See for example, Mandel, Ernst (1975), *Late Capitalism,* NLB, London.
59. The World Bank (1980), *World Development Report 1980,* Oxford University Press, New York, pp. 112–13.
60. For instance Helen Hughes and colleagues argue that official exchange rates compared to purchasing parity exchange rates result in official estimates of growth rates of income in LDCs being understated relative to those of industrialized countries. See
   (a) Hughes, Helen (1980), Achievements and objectives of industrialization, in: *Policies for Industrial Progress in Developing Countries* (eds J. Cody, H. Hughes and D. Wall), Oxford University Press.
   (b) Hughes, Helen (1980), Australian international economic perspectives, *ANZAAS Paper,* May, World Bank (mimeo).
61. For background references on Lindblom's thesis see Ch. 1, note 23.

CHAPTER THREE

Technology Policy: Options and Priorities

3.1 Introduction

Scientific effort is frequently seen to be justified on the grounds that it results in 'improved' techniques of production. What indeed are the links between science and technological change? Are they as direct and certain as is sometimes supposed? Can government support for scientific effort be justified on the grounds that scientific advance is a catalyst in technological progress? In looking at some of the choices involved in government technology policy, this chapter deals with these and other questions. It also examines, in the following order, the possible role and dangers involved in government policy designed to foster or modify steps in the technology sequences of invention, innovation, diffusion of new technology and the replacement of existing equipment. Mechanisms for the transfer of technology within countries and between countries are discussed along with some of their policy implications. Furthermore, environmental overspills from new technology and the impact of new technology on employment, on the structure of society and on working conditions are also dealt with since these raise questions for government policy.

The following are some of the specific questions covered: How relevant is science to technological development in industry? What are the main steps involved in technological change? What role is played by individuals in making inventions and is there a case for government assistance to individual inventors? How is inventiveness and R & D expenditure related to the size of firms? Should the government give priority to supporting R & D in large, small or medium firms? If inventions are to be of practical economic value they need to be followed up by (economic) innovation. What role does market competition play in stimulating innovation? Should a government interfere with market competition in an economy to promote innovation and technological progress? How important are management and marketing skills in successful innovation? To what extent should a government be involved in innovation in industry, for example by building or subsidizing demonstration plants? What factors influence the rate of diffusion of new techniques in industry and what government policies alter this rate? Replacement of existing equipment provides scope for the introduction of new techniques and improved technical efficiency. Is there a case for governments to accelerate the process of replacing equipment?

Transfer of technological ideas within a country, for example by movements of personnel between firms, between government and industry or universities, can be an important stimulus to technological change. What transfer mechanisms are significant and how can a government foster the operation of these? What mechanisms exist for the transfer of technology between nations and in particular what part do multinational companies play in this transfer? Is a host country likely to benefit from such transfer mechanisms? What types of policy choices are faced by a government in decisions about the export of domestic technology?

New technology and existing technology gives rise to environmental overspills. What types of policy options does this give a government? How does new technology affect employment, working conditions and the stability of society? What policy choices arise from these effects? Let us consider these matters.

3.2 Links between science and technology and technology sequences

The source of new technology may be (a) scientific effort, (b) trial-and-error or (c) existing technology, or a combination of (a), (b) and (c). Source (c) refers to the possibility that new technology may be developed by combining existing technologies or by building on these in simple ways. Differences of opinion exist about how important science is for the development of new technology. One extreme view (a) suggests that technology develops independently of science and another (b) claims that science is necessary for the development of new technology and that technology evolves in a definite integrated sequence from science. Intermediate positions include (c) that some technology stems in a definite integrated sequence from science and that some develops independently of science and (d) this position plus the view that there is some random over-lap between science and technology. These possibilities are represented schematically in Fig. 3.1.

If the prime purpose of science and technology policy is to develop new technologies, it is important to know the main sources of these technologies so that government assistance is not misdirected. Evidence has been found to indicate that links between basic science and technology are not very direct. For instance Parker has observed:

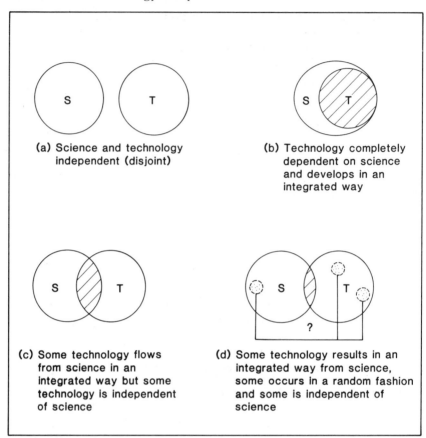

Fig 3.1 Four possible relationships between new technology and science.

In a study of publications in science and technology it was found that direct interconnection between the fields is not frequent and that technology builds on technology. Similarly in a study in the UK of Queen's Award winning developments the same conclusion was reached. Only a handful of direct connections was found, and the authors conclude 'that the great bulk of basic science bears only tenuously if at all on the operations of industry'.... It is by no means self-evident that basic scientific research provides the discoveries which industry takes up and applies. The connections between the types of activity can be indirect and complex.[1]

Nevertheless, even though the links between basic science and technology are not direct, the indirect links may be important. Advances in basic science help to add to the 'technology bank of ideas' and scientists with some training in basic science may be able to draw on this in their development work even if they do not keep up with continuing developments in basic science.[2]

Nevertheless, while education and basic scientific knowledge can play an important role in technological invention, we must be careful not to exaggerate its importance. P. Johnson has pointed out that:

> In the United States a study [by Schmookler[3]] has shown that over half of the recent inventors included in the sample were not 'college trained'. The evidence is extremely limited but it does suggest a continuing role as an inventor for the person who is not a 'trained' scientist or engineer. It is also worth pointing out that an intimate acquaintance of the preceding technical and scientific literature may not be an essential prerequisite qualification for the inventor or scientist. Banting, the discoverer of insulin, was not acquainted with all the literature that preceded his work; indeed, if he had been he might have been put off from working in the field. This of course is only one example limited to scientific investigation and it is far from providing evidence for a general case. However, it does serve as a warning against oversimplified generalizations.[4]

3.2.1 *Technology sequences*

Several stages occur in the development and use of new and successful technology. These idealized stages are represented in Fig. 3.2. *Invention* is assumed to be the first stage in technology and represents the time at which the technical possibility of a new process or product is worked out and proven. At this stage a patent application may be lodged. *Innovation* refers to the first commercial use of the technique. Before innovation occurs a considerable amount of development work is necessary to perfect the mass production or use of the invention. The work of J. Enos[5] indicated that lags of fourteen years or so are common between an invention and an innovation, but these lags can vary widely. Entrepreneurial ability and sound marketing acumen are necessary to ensure the commercial success of an innovation. Even after the innovation, the innovator may continue with development of the new technique modifying it in the light of experience from production.[6]

Once a successful innovation has occurred, the use of the technique is likely to spread. The speed at which this *diffusion* occurs in industry is likely to depend on many factors. Entrepreneurship, the profitability of the new technique, the extent to which the new technique can easily be copied or imitated, the force of patent restrictions, if any, will all play a part. After diffusion of the technique has occurred, early adopters may find that they have to *replace* equipment embodying the once new technique because (a) the equipment has worn out, or (b) adopters may find that the technique is superseded by another technique and that it is worthwhile for them to adopt the later technique, or (c) prices of resources may have altered in a way

80 Science and Technology Policy

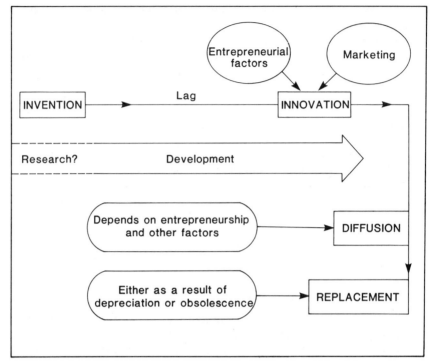

Fig. 3.2 Stages in the development and use of new technology.

which makes it profitable to use an alternative technique.

Each of the stages in the development and use of new technology is now discussed in more detail. Governments, depending upon their priorities, may wish to intervene at any of these stages to alter the development, the use and diffusion of new technology.

3.3 Inventions

While an invention is characterized in the above discussion as occurring at a precise point of time, many inventions evolve almost imperceptably with the passage of time. Some inventions are the result of the gradual accumulation of small improvements rather than a major technical or scientific breakthrough.

A number of simplistic explanations of factors determining the output of inventions exist.[7] One view asserts that inventions result from 'individual flashes of genius', another view is that they are a mechanical outcome of building on the stock of ideas and still another is that inventions are made in response to *need*. The latter view reflects the adage that 'necessity is the mother of invention'. In practice, the output of inventions may depend on all these factors. From a government policy point of view it would appear to be important to foster conditions in which individual genius can flower, to ensure the stock of ideas is maintained and extended and that society's needs are adequately signalled to inventors.

3.3.1 The role of the individual inventor

Patent statistics in the UK and the USA indicate that there has been a secular decline in the proportion of patents filed by individual inventors compared to companies in this century. This appears to support Schumpeter's and Galbraith's view that as a result of the growth of modern large companies with their own research laboratories, the importance of the individual inventor will decline.[8] This was a matter for alarm for Schumpeter as he felt that it could signal the decline of capitalism because in his view corporate research activity is less productive than individual effort.[9]

However, patent statistics take no account of the quality of inventions. To allow for this, Jewkes, Sawers and Stillerman[10] selected what they (in consultation with others) believed to be the seventy most important inventions of this century. They found that just over half were the results of the work of individual inventors. Individual inventors remain important even if their importance is declining. 'It is possible to argue that relatively few major advances come from industrial and commercial laboratories, for about half of the major inventions come from individuals, and there is evidence that universities and government research institutions produce significant numbers of major inventions.'[11]

There are a number of reasons for expecting this result. Firms tend to have short-term horizons for recouping risky investments and the development of a major invention is risky with the development period often being long. Since 'most firms seek to reduce, or avoid, uncertainty, it seems to make sense for them to concentrate on research and development that is aimed at less ambitious and more predictable results. . . . To individual inventors, and to inventors in sponsored research institutes, neither time nor uncertainty pose such severe problems'.[12]

Nevertheless, the contributions of individual inventors appear to be declining. In some cases would-be individual inventors are employed in companies. In other cases, the availability of finance has become a major restriction on their research activities. Governments need to give continuing consideration to individual inventors in their science and technology policies. They may be able to assist with grants and the encouragement of semi-independent research centres along less orthodox lines than at present.

3.3.2 Size of firms and inventiveness

There is little doubt that the major proportion of industrial R & D in industrial countries is accounted for by a few (large) firms. In the USA, UK, France and Italy the twenty firms undertaking most industrial R & D accounted for about 50% or more of industrial R & D.[13] Furthermore the research intensities of smaller firms (say those with 200 employees or less) are lower than those of larger firms on average.[14] However, the work of Mansfield[15] and others indicates that research intensities do not rise in step with firm size as a rule. The typical relationship of research intensity to the size of the firm might be like that indicated in Fig. 3.3. Maximum intensities

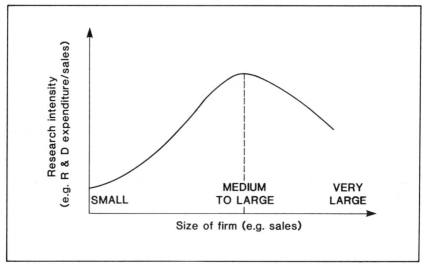

Fig. 3.3 Relationship between resource intensity and firm size. On average, research intensities appear at first to increase with the size of the firm and to decline for firms of very large size.

may occur for medium- to large-sized firms but decline for very large firms. The exact size at which the maximum intensity is likely to occur appears to vary from industry to industry. It should be noted, however, that despite declining research intensities, the largest firms in the economies for which evidence is available still undertake more R & D than smaller ones. On average total R & D appears to rise in the way indicated in Fig. 3.4.

However, R & D spending only measures research input. It is important to assess the productivity of this input in terms of inventions generated. Evidence here is difficult to interpret. Schmookler has put forward evidence to indicate that the cost per patent in terms of R & D expenditure is higher for companies in the USA with 5000 employees or more than for those with fewer employees.[16] We cannot really conclude from this that larger firms are

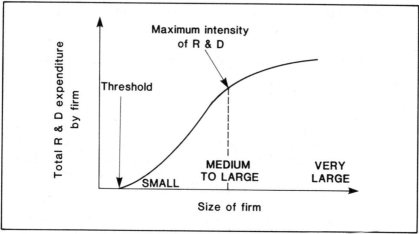

Fig. 3.4 Relationship between total R & D expenditure and firm size. Very large and large firms undertake (on average) a greater amount of R & D than medium-sized ones despite lower R & D intensities.

less efficient than smaller ones. For instance, the value of the inventions would also need to be taken into account. However, Mansfield's studies of the US chemical, steel and petroleum industries indicate that with R & D expenditure constant, inventive output tends to decline for the largest firms. He concludes 'contrary to popular belief, the inventive output per dollar of R & D expenditure in most of these cases seems to be lower in the largest firms than in large and medium-sized firms'.[17] Furthermore, there do not appear to be substantial economies of scale in R & D beyond certain limits. In some industries, the largest R & D laboratories are of a size in excess of these limits and no economies in invention can be expected from increasing their size.[18]

Economic research to date throws doubt upon the proposition that big research units are most productive of research results or inventions. In reality, there is probably some *optimum spectrum* of different sized research units in an industry or field. This needs to be taken into account in determining priorities for government support of industrial R & D.

3.4 Innovations

If R & D and inventions are to have practical value they must be followed by innovation. In the UK at least large firms (those with 1000+ employees) are responsible for most innovations. A study by Freeman[19] indicated that these firms were responsible for almost 80% of innovations in UK industries in the period 1945–70. The proportionate share of medium and large firms (200+ employees) in innovation is in excess of their proportionate share of employment and output. On the other hand, Mansfield's research into selected US industries indicates that innovating intensity may fall off for very large firms in some industries.[20] On the whole, largest firms do appear to do a proportionately larger share of the innovating.

P. Johnson sums up the results of empirical work in economics in this area as follows:

> Most industrial R & D is concentrated in large firms. Research intensities tend to increase with firm size, although there is some evidence to suggest that this relationship does not hold beyond a fairly large size. Most of the empirical work that has been done on invention and innovation indicates that there is an important but possibly declining role for the individual and small firm, and that a number of important innovations have been made through the formation of new firms. This should serve as a warning against policies that aim to increase the rate of technological change by means of greater industrial concentration. Such policies may be helpful where they enable firms to reach a threshold level of R & D spending. Beyond this level however there may be little to be gained from greater size . . . .[21]

3.4.1 *Market structure, market competition and innovation*

Schumpeter was of the view that innovation is more likely to occur under monopoly than other market forms.[22] Innovation is a risky and uncertain activity. Because a monopolist controls his market, innovation is less risky for him. The fact that a monopolist also may make above normal profits may give him a reserve against which to take risk and provides him with funds for investment in R & D. Of course, it might be put that a monopolist may prefer a quiet life to change and thrust. In response to this, Schumpeter argued that no monopolist has a watertight monopoly because he faces competition from new products in the long run.

However, intense competition does not provide fertile ground for innovation. Under intense competition in which any competitor can immediately imitate a new innovation, the innovator obtains little reward and indeed can make a loss on the innovation even though it is socially worthwhile. For instance, suppose that the widespread use of an innovation in the industry lowers the cost of production of its product and therefore *reduces the price of the product*. At the lower price, the innovator's profit may be no higher than before the innovation and if the diffusion of the invention is rapid he may fail to recoup his development costs. While the patent system is intended to help remedy this defect, its operation is far from perfect.[23]

Empirical evidence in this area is inconclusive. Examples can be given of technologically progressive monopolists and of technologically backward ones.[24] It is possible that, on the whole, some competition (actual or potential) but not intense or perfect competition is likely to be most conducive to innovation. The relationship might be like that indicated in Fig. 3.5. Oligopoly (an industry in which there are few firms) or monopoly with some barriers to the entry of new firms but not insuperable barriers could be the most innovative form of market competition.[25] Governments need to take such possibilities into account when considering their policies dealing with market competition.

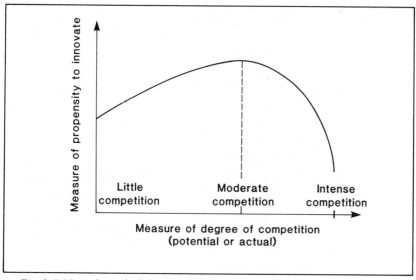

Fig. 3.5 Hypothetical relationship between the degree of market competition and the incentive or propensity to innovate. The propensity to innovate is possibly lowest under conditions of intense competition and greatest for a moderate degree of competition.

In some industries, it is possible that the government must partly play the role of an innovator. In industries consisting of many small competitive firms, any individual firm has little incentive to innovate. The government may need to develop inventions and demonstrate these to firms in the industry as is done by extension work in agriculture. Economic arguments can be advanced in favour of such intervention.

The extent to which government should involve itself in the innovative process in industry is a debatable point. In particular there has been criticism of government involvement in innovation in big science industries in which capital costs and risks have been high.[26] The ability of governments to screen projects for support has been questioned and on the whole evidence appears to indicate that negative or poor returns have been received from public funds used to support major innovations in industry. A number of countries, however, do have schemes to assist innovation in secondary industry. In the UK, for instance, financial support is provided for the development of new air engines and the government has rescued a number of innovative firms from financial difficulties.[27]

Success or failure in the innovative process is extremely dependent on management and marketing skills, as well as other factors. Evidence from the UK[28] indicates that:

> Innovation involves a complex sequence of events through from discovery up to selling. A one-variable explanation does not provide an adequate description of the process. Five statements summarize the influences which best distinguish success. These are:
> 1. successful companies have a much better understanding of user needs;
> 2. they pay much more attention to marketing;
> 3. development work is performed more efficiently but not necessarily more quickly;
> 4. more effective use is made of outside technology and scientific advice, even though successful firms perform more of their own R & D;
> 5. responsible individuals in successful innovations are usually more senior and have greater authority than their counterparts who fail.
>
> In essence failures reveal themselves by an ignorance of users' requirements; by neglect of market research publicity and user education; elimination of technical faults after launch; poor contacts with the scientific community relating to their specific area of technology; and a lower company rank of the individuals responsible for the innovation concerned.[29]

The important factors in success need to be kept in mind by managers and transmitted to would-be managers of companies.

3.5 Diffusion of new technology

The amount of social benefit received from a successful invention is likely to depend upon how widespread its use becomes and how quickly its use spreads. Therefore, the rate and extent to which innovations spread are of concern to government policy-makers. Government efforts in stimulating new technology must be balanced between the encouragement of inventions and innovation and the stimulation of the diffusion of new inventions.

It ought not to be thought that if a new technique ensures lower costs per unit of output than does existing techniques, it is socially optimal to scrap all machines incorporating past or existing techniques in favour of the new. While any additional machines to be installed should incorporate the new technique, old equipment should not necessarily be scrapped in favour of the new. Provided that the operating costs (per unit of output) using the old equipment are less than the operating costs plus capital charges (interest or normal profit) per unit of output using the new equipment, the old equipment should be retained. If the opposite is the case the old equipment should be scrapped in favour of the new.[30] Note that in the decision, the capital costs of installed equipment are not taken into account. It is a sunk cost. Just how many old machines it pays to scrap depends upon the vintages of installed machines, that is on the age distribution of the population of installed machines. Where the majority of machines are of the oldest vintage, a large proportion of the machines in the industry may soon be profitably replaced by new ones incorporating the latest invention, but where all machines are of recent vintage the diffusion process may be slow and justifiably so,[31] as is illustrated by Fig. 3.6.

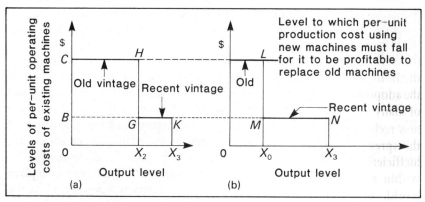

Fig. 3.6 The age (vintage) distribution of existing machines determines the proportion of machines that it is optimal to replace when technological progress occurs. Replacement is only optimal when *total* per unit production costs with new machines is less than per unit *operating* costs using the old. Thus if per unit production cost using new machines is between OB and OC, it is profitable to replace few existing machines if the distribution in Fig. 3.6(b) applies but a high proportion if 3.6(a) applies.

Nevertheless, the diffusion of new techniques may occur at a slower rate than is most profitable and a slower rate than is socially optimal. This may occur because of the following factors:

(a) ignorance or lack of knowledge about the availability of the new technique;
(b) managerial inertia (sleeping managers);
(c) uncertainty about future demand for products to be produced using the technique;
(d) technological uncertainty;
(e) labour fears and trade union resistance.

Where markets fail to provide communication government may be able to assist in improving information flows. It may also be able to assist in alleviating some of the other impediments mentioned.

Empirical studies of the diffusion of new techniques indicate that the relative profitability of new techniques is a major determinant of their rate of adoption and diffusion.[32] Mansfield's work[33] indicates that techniques involving little capital investment and high profitability can be expected to spread quickly. The probability of adoption of a technique by a firm in an industry appears to rise with (a) the proportion of other firms using the technique, (b) the expected profitability of the technique and (c) the smaller the needed investment. However, rates of diffusion seem to differ from industry to industry.

Competition may also be a factor in the spread of new techniques. In industries which are competitive, market competition provides an external pressure for the adoption of cost-reducing techniques. As new cost-reducing techniques are adopted by more firms in a competitive industry, the price of the product of the industry falls and this places non-adopters in the industry under pressure if the new technique is more profitable for them. In the end, the survival of firms of non-adopters may depend upon the adoption of the new technique. However, competition and the existence of many small firms in an industry is not a guarantee of rapid diffusion of new techniques. If all firms in the industry tend to be insular and backward, the spread of techniques can be slow as Leibenstein's work on the 'X-inefficiency' of firms indicates.[34] The competition of firms may take place within a limited framework and a common restricted perception of the world, and yet such competition can be intense within this limited framework.

Various simple diffusion models have been developed to predict the path of the spread of new techniques and new products. Since these are well summarized by Kotler[35], I shall not outline these in detail but will comment briefly.

Diffusion (cumulative diffusion) seems most commonly to follow an elongated 'S' form (logistic form) as in Fig. 3.7. The likelihood of this type of curve applying may be argued on the basis that the relative frequency distribution of potential adopters of a new technique or product follows a normal form or a bell-shaped form as in Fig. 3.8, in which the degree of innovativeness of a population is normally distributed.

The likelihood of such a curve can also be argued on the basis of epidemiological models. These use the principle that the spread of information about new techniques depends upon the frequency of contacts. Ozga discusses the application of these models to the spread of new products and new techniques in some depth.[36] A further factor making for an S-shaped cumulative diffusion curve is the S-shaped individual learning curve.

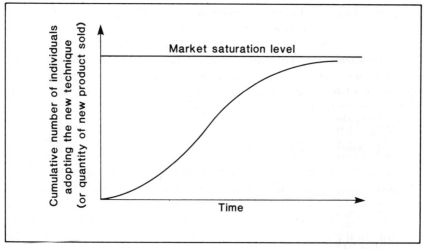

Fig. 3.7 The diffusion of new techniques and products. This may typically follow an S-shaped cumulative form.

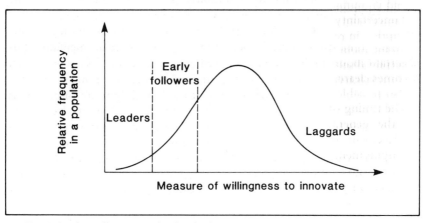

Fig. 3.8 The willingness of individuals to innovate and adopt new techniques and products may be normally distributed in a population. This could be one factor resulting in an S-type cumulative adoption curve.

Government policy-makers face at least one dilemma in innovation and diffusion policy. The patent system illustrates this. The patent system rewards innovators by providing them with a monopoly on their invention. To maximize profit from their invention inventors may need to charge a price for its use which considerably slows down the rate of spread of the technique.[37] In France a novel approach to dealing with this dilemma has been tried. Income earned from licence fees are a tax deduction.[38]

Governments may also be able to play a positive role in promoting diffusion of new techniques in other ways. In Britain, for instance, the government has made new equipment available to some industries on loan with an option to purchase after a specified period of time.[39] As is done in many countries, governments can assist diffusion as in agriculture by arranging for field days and demonstrations of new techniques and testing these in new localities. In some countries, e.g. Japan, a similar system operates for small businesses not in agriculture.

3.6 **Replacement of equipment**

To some extent the economics of replacing equipment has been touched upon in discussing the diffusion of new technology. Even where there is no technological progress in an industry it is possible that equipment is not replaced quickly enough or in a socially optimal pattern. Nevertheless, let me reiterate that wide differences in productivity in industry can be optimal.[40]

Old equipment may be retained too long because of inertia, ignorance and uncertainty. The installation of replacement equipment involves some disruption in regular or habitual operations of the firm, managers may be ignorant about the level of operating costs of old equipment or they may be uncertain about future sales of their product and defer replacement until this becomes clearer. In the latter case deferment could be socially optimal but it is also possible that collectively the uncertainty may be unjustified.

The timing of replacement investment can have important implications for the general level of employment and demand in an economy. Replacement may be stimulated by governments during periods of unemployment and low levels of demand to raise employment. In some countries, governments have given companies extra investment allowances for taxation purposes and have permitted accelerated depreciation for a limited period in order to stimulate replacement investment at a propitious time.

A government may also wish to interfere with the free replacement pattern of equipment for environmental reasons. As equipment grows older it is possible that the deleterious pollution overspills from it may increase in a way which makes it socially optimal to replace it before its commercial life is finished. Governments may affect such replacement by imposing a special tax on old equipment or providing a subsidy for its replacement. The case for interfering with the commercial pattern of replacement is likely to be even stronger if technological progress is occurring and is resulting in the supply of less polluting equipment.

3.7 Domestic technology transfer

Evidence indicates that most of the ideas used in innovations by firms come from outside their firm. A study by Langrish and coworkers of the 51 innovations in the UK that received Queen's Awards to Industry in the 1960s revealed that 102 of the 158 most important ideas used in these innovations came from outside the award-winning firm. The method of transfer of these important ideas is indicated in Table 3.1. The Table indicates that transfer of individuals between organizations is the greatest single source of transfer. Education plus literature accounted for 18.5, the government for 6 and conferences for 2.5 important ideas making a total of 25%, and visits overseas were not unimportant.

Table 3.1 Method of transfer of 102 important ideas used in award-winning innovations and obtained from outside the firms winning the Queens Award for Industry in UK.

Transfer via person joining the firm		20.5
Common knowledge via	industrial experience	15.0
	education	9.0
Commercial agreement (inc. takeover and sale of know-how)		10.5
Literature (technical, scientific and patent)		9.5
Personal contact in UK		8.5
Collaboration with	supplier	7.0
	customer	5.0
Visit overseas		6.5
Passed on by government organization		6.0
Conference in UK		2.5
Consultancy		2.0
	Total	102.0

Note 0.5 indicates that some sources are not mutually exclusive.
Source Based on J. Langrish *et al.* (1972), *Wealth from Knowledge*, Macmillan, London, Table 7.

The 'Bolton' Committee of Inquiry into Small Firms in the UK suggested that important factors in the transfer of ideas are movements of people, mergers and takeovers. The Committee held:

> Possibly the majority of innovations ... arise from a little understood process of movement of people: through small firms being acquired by larger firms or licensing larger firms or acquiring licences from them; by staff leaving one company and joining another and also through the so-called spin-off process where an individual leaves a large company, university or laboratory (some-

times because he cannot get his ideas implemented) to set up his own business, perhaps to succeed spectacularly, perhaps to fail and possibly in either event, to sell out to a larger firm.[44]

Transfers of technology may be affected by transfers of individuals from public institutions (universities and government bodies) to private institutions, transfers of individuals between private institutions and less commonly by transfers of individuals from private to public institutions. Not infrequently individuals leave employment in public and private institutions to commence their own businesses to make use of ideas which they have developed in their previous employment. In the case of inventions made in the public sector, it is as a rule essential for these to be transferred to private industry if they are to be used. The only effective way of transferring some of these may be by the movement of individuals from the public sector.

Complex issues can be involved in transferring inventions from the public sector to the private sector, especially when these inventions are of value to secondary rather than primary industry. In general, it is desirable for public bodies to patent their inventions, but the use of their patent should be in the public interest. If a public body fails to patent an invention, a private company may find out about the invention, patent it in its own name and restrict the use of the invention. Considerable costs and risks can be involved for companies in bringing an invention of industrial value to the market place (development costs, tooling costs and marketing costs of the typical pattern indicated in Table 3.2 must be incurred) and this sometimes may require the government to license the invention exclusively, at least for a period of time, to one industrial firm or at the most a limited number. These problems are covered in some detail in my paper. 'Patenting and licensing of government inventions – general issues raised by Australian policy'.[45]

The government may be able to assist in the transfer of ideas within the economy by facilitating transfers of employment and by encouraging temporary interchange of staff. Portable superannuation schemes and more flexibility in operating secondment and exchange schemes are needed. This is not to say that caution is not required in some cases where a possible transfer might not be in the public interest, e.g. in the defence field, or the transfer might occur prematurely before a public body can patent an invention.

In Britain, and some other countries, the government has set up a public corporation to facilitate the patenting and use of inventions from the public sector and to foster use of inventions which otherwise might not be used. While the experience of the British National Research and Development Corporation (NRDC) is that there is not a large untapped supply of individual inventions of commercial potential waiting to be applied, the NRDC

Table 3.2 Proportionate division of costs in bringing an invention to market: USA estimates.

Stage		Average % of total cost arising at each stage	
		I	II
1 Applied research	} Invention costs	9.5 } 17.1	5–10
2 Specifications		7.6	
3 Prototype or pilot-plant	} Development costs	29.1	10–20
4 Tooling and manufacturing facilities		36.9	40–60
5 Manufacturing start-up		9.1	5–15
6 Marketing start-up		7.7	10–25

Sources I: E. Mansfield *et al.* (1968), *Research and Innovation in the Modern Corporation*, Norton, New York. II: US Department of Commerce (1967), *Technological Innovations*.

has been able to bring a number of Research Council and university inventions into profitable commercial use. While the Corporation made a net commercial loss overall in the beginning, it now operates at a profit and it is possible that the operations of NRDC yield a positive social benefit.[46] The desirability of such a body, especially as far as university and government inventions are concerned, is worthy of consideration in countries where similar bodies are absent.

3.8 International transfer of technology

National frontiers may provide resistance to the free flow of technology. But the mechanisms for transferring technology internationally are similar to those within a nation. As in national market, new technology may be embodied in products, and merchants, distributors or salesmen may be the most important single point of contact with potential customers and the main agents for diffusion of the new technology, through sales of products incorporating it. Griliches found that in the USA, seed merchants were the main agents spreading the use of hybrid corn by farmers[47] and salesmen for instance, play a large role in spread of use of new drugs by medicos. International marketing networks and distributors play a large role in the diffusion of technology internationally.

National companies may first come to know of new foreign technology in a variety of ways, apart from possibly seeing it being used by other companies in the economy. They may see the product incorporating the technology advertised in an international magazine; they may learn about it from a salesman; it may be brought to their attention by an employee from the foreign country where the invention is used;[48] managers may see it on a visit to the overseas country when the technology is being used or learn about it at a national or international conference or it might be brought to their attention by a government department or consultancy. Empirical work is needed to determine the relative importance of these different channels of communication.

3.8.1 Multinational companies

Multinational companies are important vectors in the international diffusion of technology. Parker has said in relation to this:

> The multinational is capable of reducing the separation of markets. It is thus likely to be important in the diffusion of technology. These companies have a global spread of subsidiaries which are co-ordinated through head office. Where headquarters are in the technologically advanced nations, this suggests a potential to unify and accelerate the rate of spread of innovations. Adoption rates normally vary between nations. The example of plastics illustrates the wide range of times taken to spread between different countries. However, the foreign subsidiary has a technological capability which is effectively independent of its local environment. Its skill and knowledge base is derived from the parent company, not from the host economy. This is the technological independence effect. Consequently the rate of adoption of innovations is likely to reflect not the attitudes prevalent in overseas locations, but those in head office. Impetus and timing may be dictated by headquarters, and is unlikely to reflect some 'natural' process of diffusion. Knowledge of innovations is unlikely to be much affected by geographical separation from the home environment. Intra-company movement of personnel should reduce knowledge barriers. Percolation of ideas and products will be hustled by fiat, and will not necessarily reflect local market conditions. The rate at which innovations are taken up will probably reflect parent company attitudes, and is likely to be very different from that typical in the host economy. Diffusion rates may therefore approximate to those typical at home. Where this is an advanced industrial nation like the USA the rate of spread will presumably be faster.[49]

One theory used to explain the prevalence of multinational producing enterprises is that multinational operations are needed by innovative firms if they are to capture or appropriate a substantial share of the benefits from their innovations.[50] Multinational firms tend to be in industries which are R & D intensive and innovative. If such companies do not produce overseas when substantial markets for their products exist there, they run the risk that other firms will enter these markets and imitate their inventions. Patents may be a poor protection against such imitation because infringements of patents have to be detected, the company must have applied for patents in the country concerned and legal costs in disputed cases can be high.

It might be thought that licensing would be a satisfactory alternative to production by a subsidiary of a multinational innovator. This is not always so from the parent company's point of view. It may give prestige to the licenser, may enable the licenser to develop other competing inventions and may make it more difficult for the licensee to enter the market subsequently.

A country may benefit from the presence of multinationals by a speedier diffusion of technology from abroad and national employees of the company may learn skills which subsequently enable them to establish research-intensive firms of their own. On the other hand, possible disadvantages of such companies are that they may (a) make it more difficult for infant domestic research-intensive firms to grow to maturity[51], (b) lead to the export of inventions developed in the home country on conditions not to the advantage of the home country[52] and (c) lead to *innovations* based upon domestic inventions occurring first in economies abroad.[53] This (c) may deprive the domestic economy of a favourable position in the product cycle, as far as some products are concerned. Considerations of this kind have clearly weighed heavily in Japanese decisions to restrict foreign investment and in Canadian policy in giving preference to domestically owned firms in government R & D grants.

Some writers have also expressed doubts on other grounds about the benefits to small host countries of relying on multinational or transnational companies as vehicles of technology transfer. It is pointed out that (a) in small host economies such companies could become politically influential (as in the Chilean case) thereby reducing the political independence of the host, (b) such companies have the means to transfer income abroad to tax havens or low-tax countries thereby reducing their tax liability in the host country, (c) their international monetary movements may in certain circumstances destabilize the exchange rate of the host country and (d) in less developed countries such companies may transfer inappropriate technology and create ghetto or insular-type development employing few (relying on expatriates and overseas trained management) and providing little technology feedback to the bulk of the economy.

The importance and desirability of different international transfer mechanisms may differ from country to country and may depend upon the stage of economic development of a country. There is a need for further study of these mechanisms especially for dependent small economies.

3.8.2 *Export of domestic technology*

The export of a country's technology is likely to be facilitated on favourable terms when it is the *headquarters* for a number of multinational companies.

96 *Science and Technology Policy*

Such companies appear to be efficient in extracting benefits from innovation and invention. Large companies with multinational operations are possibly in the best position to appropriate most gains from the R & D effort of a nation. To the extent that a small country is without such companies and also has a small home market it may be uneconomic for it to enter early in the product cycle and it is at a disadvantage in the 'international R & D stakes'.

Nevertheless, even the competitive industries of a country can sometimes benefit by the export of domestic technology. 'A classic case' is Australia's export and promotion of wool textile technology through the International Wool Secretariat to promote Australian wool exports.[54]

There are many issues to be explored in the export of technology. To what extent is government interference, like that which has occurred in Japan, beneficial? When does it pay a country to try to keep its technology secret and/or not export it? How can a country or its firms obtain the best conditions in exporting or licensing its (their) technology? Unfortunately these are questions which cannot be explored here.

3.9 Environmental overspills and technology

Some of the policy problems that occur when technology gives rise to environmental overspills or externalities have already been discussed in Chapter 2. As pointed out, markets fail to guide firms adequately in the appropriate social choice of technology when overspills are important. Government interference with free choice may be called for. Apart from encouraging the development of technology which reduces environmental overspills, governments may need to control the technology used by individuals and firms and the way in which it is used.

A number of policies are available to governments. A government may (a) tax emissions of the offending pollution and this will provide an incentive to economic agents to reduce the pollution and choose techniques to reduce it, (b) promulgate regulations limiting allowable emissions, (c) ban or place special taxes on technology which has very unfavourable externalities (e.g. cars without mufflers) and (d) provide subsidies for the introduction of pollution-reducing technology. It is not possible to debate the merits and drawbacks of these various approaches here.[55] My main concern is to point out that positive government action may be called for when different types of technology have environmental overspills.

Nevertheless, it should be noted that even when the socially optimal available technology is employed optimally, pollution and environmental overspills may still exist. As a rule, it is not optimal to eliminate all unfavourable environmental overspills.[56] The added cost of reducing overspills must be weighted against the extra social benefit. Once pollution is reduced to a level where the extra cost of further reducing it exceeds the extra social benefit, no further reduction in pollution is called for.

Technology having uncertain long-term environmental effects and which could possibly have catastrophic consequences for human existence poses the most difficult choice of problems for governments in practice. While some see the nuclear generation of electricity as particularly hazardous, others see its possible environmental impact as less risky. Nevertheless, there is uncertainty and differing opinions about the degree of risk which society should take in using such technology and this makes the choice for governments difficult. Given the risks, the priority placed on the use of nuclear technology will depend to some extent on the relative costs of generating electricity by other means. Important moral decisions, however, cannot be avoided in developing and using a whole range of new technology.

3.10 Employment problems and other social aspects of technology

Technology has transformed the structure of our society. It has resulted in a long-term shift of population from agriculture to secondary industry to tertiary industry, and an accompanying rise in urbanization, has made increased leisure time available to us, given us means to enjoy leisure more fully, and has possibly been a powerful secular force bringing about greater equality of income.[58] Furthermore, new technology is continuing to make further changes in society as a result of automation and the growing use of computers. Whereas earlier technological revolutions tended to reduce the need for unskilled labour in difficult tasks, the information-technology revolution is replacing the need for clerical workers and those offering basic (primitive) intellectual skills.

3.10.1 *Employment*

The question which immediately arises is whether such technological developments can result in permanent unemployment of groups in society.[59] In my view this can occur if minimum wage rates are rigid. If the demand for unskilled and semi-skilled labour falls substantially as a result of technological progress and minimum wage rates for these groups are maintained, unemployment amongst these groups can be expected to increase. Schemes to raise wages so that these groups share in the increased productivity may in fact increase the degree of unemployment. It might be thought that such technological unemployment can be overcome by moving members of the unemployed group to other positions. However, they may need retraining. While this is not an insuperable obstacle, it is possible that this group is incapable of filling the type of highly skilled positions in demand because of their attitudes to education or the fact that they do not have intellectual-type strengths and abilities. While new technology creates new opportunities (e.g. in its design, maintenance etc.), it is not likely to create these opportunities evenly for the population. There is a risk that a high proportion of the less skilled may become permanently unemployed (given rigid wage rates) and may not share in the fruits of technological progress, unless of course they are employed by governments in make-work schemes. We may reach a situation in which there is a permanently unemployed group in society (supported by government unemployment benefits) and a well-off employed group. If the present situation is of this type, many of the schemes being put forward to rectify it must fail. For instance, early retirement and a shorter working week might only have a marginal effect. These schemes may reduce demand, and are unlikely to generate employment to the extent of employment displaced. For instance, consider early retirement. A proportion of those retiring early will be highly skilled and need to be replaced by skilled labour, but if unskilled labour cannot be substituted for skilled and if skilled labour is fully employed, a proportion of places arising from early retirement cannot be filled. In the *extreme* case, net employment may decline. For example, suppose that one skilled person is required to supervise a fixed number of unskilled. The employment of the unskilled if skilled individuals are in short supply will depend on the number of skilled individuals employed. *An early retirement scheme which reduces the available supply of skilled individuals reduces the employment of the unskilled.*[60] The supply of skilled and *talented* labourers is the constraint in this model. Early retirement may in addition lead to a fall in demand and can have serious sociological effects unless individuals are prepared for it.

Technological change has differential impact on different groups in society. To some extent these effects can be compensated for by economic adjustment of the groups affected. Firms may adjust by switching from declining industries to technologically progressive and expanding industries and workers may do likewise. In many cases this requires leaving the region where they are at present situated. In many countries governments have provided assistance to facilitate such adjustment.

However, in some cases, policy has been directed not at increasing mobility of economic agents between regions but at transforming economic activities in regions in which established industries are declining. Attempts have been made to foster new industries and upgrade technology as in the UK and many other OECD countries.[61]

In a technologically changing society the nature of education and the possibilities for retraining are important. Education needs to provide the individual with flexibility and opportunities must exist for upgrading and modifying skills and knowledge as technological progress occurs.

It might be noted that some economists believe that clustering of technological change plays a major role in long-term cycles of business activity and general employment. Major innovative periods seem to come in bunches and these generate considerable levels of investment and expand investment at the time and are followed by lulls which correspond to relative slumps in economic activity. According to Krondratieff, cycles generated in this way are very long – of 48 to 60 years in duration.[62]

3.10.2 Working conditions

Technology has considerable impact on the work environment. Modern technology has reduced many of the unpleasant conditions of work (hard physical exertion for instance) but has not eliminated all hazards and unpleasant aspects of employment. In some cases new hazards have arisen such as in higher risks of cancer. In other cases, risks of deafness induced by noisy machinery continues. Nervous tension may have increased in the workplace (Otto Neuloh believes this is important and requires longer holiday periods to enable 'recuperation' to occur[63]), alienation may be a continuing problem and mobility and other stresses mentioned by Toffler in *Future Shock*[64] with their implications for family life and the stability of society cannot be swept to one side but must be weighed against benefits from technological change.

Unbridled technological and economic change can have sociological costs for society too high to pay for greater material welfare. Governments need to keep such considerations in mind and may need to interfere at times to moderate or cushion the speed of economic change.[65] To some extent, however, governments are likely to be constrained by the international competitive position in the extent to which they can restrict or limit technological change in domestic industries. Industries in countries without restrictions or with fewer restrictions may out-compete domestic industries restricted in their technology. In some cases it may be possible to obtain international agreement limiting technology (for instance restricting noise levels in certain types of machinery such as aircraft) and in other cases it may be worthwhile for a nation to go it alone if necessary. Sweden for instance has done the latter by introducing lower permissible noise levels in industry than in most other countries.

Industrial relations not only affect the introduction of new technology but may also alter such relations. Some systems of industrial relations slow down the introduction of new technology and can be a cause of rigidity and economic inefficiency. To discuss this here, however, would take me beyond the limits of the present monograph and would not do justice to the topic.[66].

3.11 Observations

Several of the complex steps (invention, innovation, diffusion and replacement of technology) in the technological cycle have been discussed. We have observed that new technology affects and can change the whole structure of society and that while it has brought great benefits to mankind, it is not always an unmixed blessing. It can have deleterious social and environmental effects and these must be balanced against gains, and technology modified in appropriate instances. It was observed that there is need for further study of the transfer mechanisms of new technologies, especially in and to small economies, like those of Australia. The role of multinational companies in the international transfer of technology needs more consideration especially from the point of view of dependent host countries.

The Luddite fear of technological unemployment cannot be dismissed as a fantasy because technological change usually generates structural change in the pattern of production in economies. Industries may rise and fall with such change necessitating variations in the employment of labour, variations which can take considerable time to work themselves out. The skilled may find that their skills are replaced by machines and in a world in which wage rates are relatively inflexible may find that long delays are involved in securing alternative employment.[67] New technology is likely to be a risk to the security of some of the employed. Should a government interfere to try and slow down technological change to reduce these risks or should it try to assist the displaced to find employment elsewhere? In the last case it has to be considered whether employment is available elsewhere given the institutional restraints on wage levels. Should governments bend to the discipline of international competition resulting from technological change in other countries by promoting greater technological change at home? In doing so, what regard should be paid to the environmental consequences of the new technology? Is it legitimate to ignore the globally damaging consequences of new technology (for example supersonic transport) in the interest of selfish national gain?

This raises the fundamental point of whether technological change has been made a scapegoat for ills in our society which arise principally from our desire for economic growth and our attempt to use economic growth as the touchstone for dealing with economic ills such as unemployment, poverty and an inequitable distribution of income.[68] Economic growth and technological change designed to foster it may provide short-term relief for these problems but in so doing worsen the long-term economic, environmental and social conditions faced by mankind.[69] Can we expect governments to take a long-term or short-term view, a national or an international standpoint? We cannot unfortunately rule out the possibility of short-term national interests prevailing. However, let us examine the priorities which have emerged in government science and technology policies in selected OECD countries before drawing any further conclusions on this issue.

Notes and references

1. Parker, J. E. S. (1974), *The Economics of Innovation*, Longman, London, p. 22.
2. In the case of Table 3.1, it is still possible that individual inventors relied on a good deal of basic science.
3. Schmookler, J. (1957), Inventors past and present, *Review of Economics and Statistics*, **39**, 321–33.
4. Johnson, P. S. (1975), *The Economics of Invention and Innovation*, Martin Robertson, London, p. 63.
5. Enos, J. (1963), Invention and innovation in the petroleum refining industry, in: *The Rate and Direction of Inventive Activity*, National Bureau of Economic Research, Princeton University Press, Princeton.
6. Invention may be a more continuous process than indicated in Fig. 3.2. Note also that the importance of learning by doing for costs and productivity has been emphasized in a number of economic contributions. See, for example:
    (a) Hartley, K. (1965), The learning curve and its application to the aircraft industry, *Journal of Industrial Economics*, **13**, 122–8.
    (b) Hay, D. A. and Morris, D. J., (1979), *Industrial Economics*, Oxford University Press, Oxford, pp. 49–50 and elsewhere.
7. For a summary of these see, for instance:
    (a) Johnson, P. (ref. 4) Ch. 2.
    (b) Parker, J. (ref. 1) Ch. 3.
    (c) Rosenberg, N. (1974), Science, invention and economic growth, *The Economic Journal*, **84**, 90–108.
8. Galbraith, J. K. (1967), *The New Industrial State*, Hamish Hamilton, London.
9. Schumpeter, J. A. (1942) *Capitalism, Socialism and Democracy*, 2nd edn, Harper Brothers, New York.
10. Jewkes, J., Sawers, D. and Stillerman, R. (1969), *The Sources of Invention*, 2nd edition, Macmillan, London.
11. Norris, K., and Vaizey, J. (1973), *The Economics of Research and Technology*, Allen and Unwin, London, p. 39.
12. Norris, K. and Vaizey, J (see ref. 11) p. 39.
13. OECD (1967), *The Overall Level and Structure of R & D Effort in OECD Member Countries*, OECD, Paris.
14. Cox, J. G. (1971), *Scientific and Engineering Manpower and Research in Small Firms*, HMSO, London.
15. Mansfield, E. (1968), *The Economics of Technological Change*, Norton, New York.
16. Schmookler, J. (1966), *Invention and Innovation*, Harvard University Press, Cambridge, Mass.

17. Mansfield, E. (1969), *Industrial Research and Technological Innovation,* Longmans, London, p. 42.
18. Mansfield, E. (see ref. 17) p. 42.
19. Freeman, C. (1971), *The Role of Small Firms in Innovation in the U.K. Since 1945,* HMSO, London.
20. Mansfield, E. (see ref. 17) pp. 40–43.
21. Johnson, P. (see ref. 4) p. 69.
22. Schumpeter, (see ref. 9). But as pointed out in the previous chapter, Schumpeter believed that increasing concentration of research in large industrial organizations would eventually slow the pace of technological progress.
23. In this regard see, for instance, Taylor, C. T. and Silbertson, Z. A. (1973), *The Economic Impact of Patents,* Cambridge University Press, Cambridge.
24. Johnson, P. (see ref. 4) Ch. 4.
25. (a) Johnson, P. (see ref. 4) Sections 4.3 and 4.4.
    (b) Parker, J. (see ref. 1) pp. 62–78.
26. (a) Eads, G. and Nelson, R. (1971), Government support of advanced civilian technology, *Public Policy,* **19**, 405–27.
    (b) Pavitt, K. (1976), The choice of targets and instruments for government support of scientific research, *The Economics of Industrial Subsidies* (ed. A. Whiting), HMSO, London, pp. 113–38, especially pp. 125, 126.
    (c) But see also the more general study by Pavitt, K. and Walker, W. (1976), Government policies towards industrial innovation: a review, *Research Policy,* **5**, 11–97.
27. Whiting, A. (see ref. 26b) pp. 14–18.
28. Langrish, J., Gibbons, M., Evans, W. G. and Jevons, F. R. (1972), *Wealth From Knowledge,* Macmillan, London.
29. Parker, J. E. S. (see ref. 1) pp. 83, 84.
30. For a clear account of the process of replacement see Norris, K. and Vaizey, J. (ref. 11) Ch. 7.
31. Using the Salter-type model, as outlined by Norris, K and Vaizey, J. (ref. 11) considerable replacement occurs of machines when the per unit cost of new machines falls below OC (but not OB) if the distribution of existing machines is as in Fig. 3.6 (a) but little replacement occurs if the distribution is as in Fig. 3.6 (b).
32. (a) Griliches, Z. (1957), Hybrid corn: an exploration in the economics of technological change, *Econometrica,* **25**, 501.
    (b) Mansfield, E. (see ref. 17) Ch. 8.
    (c) Mansfield, E. (1961), *Econometrica,* **29**, 741–66.
    (d) See also Ray, G. F. (1969), The diffusion of new technology: a study of ten processes in nine industries, *National Institute Economic Review,* 40–83.
33. Mansfield, E., Rapoport, J., Wagner, S. and Hamburger, M. (1971), *Research and Innovation in the Modern Corporation,* Norton, New York.
34. Leibenstein, H. (1966), Allocative efficiency vs. X-efficiency, *American Economic Review,* **56**, 397–415.
35. Kotler, P. (1974), *Marketing Decision Making,* Holt Rinehart and Winston, London, Ch. 17.
36. Ozga, S. (1960), Imperfect markets through lack of knowledge, *Quarterly*

Journal of Economics, **74**, 29–52.
37. Taylor, C. T. and Silbertson, Z. A. (see ref. 23).
38. See Whiting, A. (1976), Overseas experience in the use of industrial subsidies, *The Economics of Industrial Subsidies* (ed. A. Whiting), HMSO, London, pp. 45–63 especially p. 61.
39. Field, G. M. and Hills, P. V. (1976), The administration of industrial subsidies, in: (ref. 38) pp. 1–22, especially p. 19.
40. The average productivity in an industry may be as little as half that of productivity in best-practice plants of most recent vintage. See for instance Salter, W. (1969), *Productivity and Technological Change,* 2nd edn, Cambridge University Press, Cambridge. As illustrated in note 31, this does not imply that the industry is economically inefficient.
41. Whiting, A. (see ref. 38) p. 58.
42. Japan permits tax deferment for investment in pollution control equipment and France allows accelerated depreciation.
43. Langrish, J. *et al.* (see ref. 28).
44. *Report of the Committee of Inquiry into Small Firms* (1972), HMSO, London, p. 53.
45. Tisdell, C. (1974), Patenting and licensing of government inventions – general issues raised by Australian policy, *Australian Economic Papers,* **13**, 188–207.
46. Johnson, P. (see ref. 4) Ch. 6.
47. Griliches, Z (see ref. 32a)
48. At a conference some manager-owners of engineering firms in Newcastle indicated to me that they visit the United States annually to look for new innovations and adapt these to Australian conditions. Proceedings of the conference are in Davies, M. *et al.* (eds) (1978), *Regional Innovation and Economic Adjustment,* The Institute of Industrial Economics, University of Newcastle.
49. Parker, J. (see ref. 1) p. 206.
50. For an exposition of this theory see Parker, J. (ref. 1) p. 178, *et seq.* For a more recent work bringing together the main streams of thought about multinationals, see Hood, N. and Young, S. (1979), *The Economics of Multinational Enterprise,* Longman, London.
51. See Tisdell, C. (1977), Research and development services, in: *Economics of Australian Service Sector* (ed. K. Tucker), Croom Helm, London, Ch. 8.
52. The invention may be made in a subsidiary and transferred to the parent company for little or no consideration.
53. Subsidiaries of multinational companies may speed up fast reverse-engineering by the parent company by making information available to the parent on promising technical developments in the host country. This may make it difficult for domestic firms to get a 'head start'. Inventions made by a subsidiary may be transferred to the parent company to use these first because this is most economic from the multinational company's point of view. The host country's followership status is thereby maintained.
54. In this regard see for instance Tisdell, C. A. and McDonald, P. (1979), *Economics of Fibre Markets: A Global View of the Interdependence Between Man-made Fibres, Wool and Cotton,* Pergamon Press, Oxford, Ch. 2.
55. For a discussion of these see, for example, Pearce, D. W. (1976), *Environmental Economics,* Longman, London.

56. See, for instance, Tisdell, C. A. (1974), *Economics of Markets: An Introduction to Economic Analysis*, Wiley, Sydney, Ch. 15.
57. For a review of these hazards see:
    (a) Baumol, W. J. and Oates, W. E. (1979), *Economics, Environmental Policy, and the Quality of Life*, Prentice-Hall, Englewood Cliffs, Ch. 3
    (b) For special attention to the nuclear hazard and associated choice and moral questions, see Shrader-Frechette, K. S. (1980), *Nuclear Power and Public Policy*, Reidel, Dordrecht, Holland.
    (c) Also revelant is Tisdell, C. A. (1979), Scientific and technological risk-taking and public policy, *Theory of Knowledge and Science Policy* (eds W. Callebaut *et al.*) University of Ghent, Belgium, pp. 576–85.
58. (a) For a general discussion of some of these issues see Kendrick, J. W. (1964), The gains and losses from technological change, *Journal of Farm Economics*, **46**, 1065–72.
    (b) Robertson, D. J. (1964–65), Economic effects of technological change, *Scottish Journal of Political Economy*, 11–12, 180–94.
59. In the light of the current recession there has been renewed debate about the possibility of technological unemployment. To some extent this recent debate is reminiscent of the 1930s. See, for instance:
    (a) Kaldor, N. (1932), A case against technical progress, *Economica*, **12**, 180–96.
    (b) Staehle, H. (1939–40), Employment in relation to technical progress, *Review of Economic Statistics*, **21–22**, 94–100.
    More recent discussions include:
    (c) Robertson, D. J. (ref. 58b)
    (d) Stieber, J. (ed.) (1966), *Employment Problems of Automation and Advanced Technology*, Macmillan, London.
    (e) Stoneman, P. A. (1976), *Technological Diffusion and the Computer Revolution*, Cambridge University Press, Cambridge.
    (f) See also some of the contributions in Aislabie, C. J. and Tisdell C. A. (eds), (1979), *The Economics of Structural Change and Adjustment*, Institute of Industrial Economics, University of Newcastle. The contribution of P. Harris is of particular relevance. Furthermore, see note 67 below.
60. In this *extreme* and possibly fanciful model the number of unskilled people employed, U, is a linear function of the number of skilled individuals, S, employed. If λ is the 'span of control' of a skilled individual, $U = \lambda S$. The upper limit to S places a constraint on the potential for employing unskilled individuals. Let $\bar{S}$ represent the upper potential for S in a population and $\bar{U}$ represent the residual pool of those who cannot achieve skilled status. Fundamental long-term unemployment of the unskilled would appear to be unavoidable if $\bar{U} > \lambda \bar{S}$. Even if all those with skill potential are brought into skilled employment, a pool of unskilled cannot be employed given existing technology. If $\bar{U} < \lambda \bar{S}$ no problem exists. Those with skills or skill potential can work in unskilled occupations. They have a flexibility denied to those not possessing skill abilities. A rise in λ in this model increases the likelihood of long-term unemployment of those without skill potential.
61. Whiting, A (see ref. 38)
62. See Garvy, G. (1953), Krondratieff's theory of long cycles, in: *Readings in*

Business Cycles and National Income (eds A. Hanson and R. Clemence), Allen and Unwin, London, Ch. 31.
63. Neuloh, O. (1966), A new definition of work and leisure under advanced technology, in: ref. 59d, Ch. 11.
64. Toffler, A. (1971), *Future Shock,* Pan, London.
65. Committee to Advise on Policies for Manufacturing Industry (1975), *Policies for the Development of Manufacturing Industry,* **1**, Report, AGPS, Canberra.
66. For further discussion of some of these matters, see Hunter L. C. *et al.* (1970), *Labour Problems of Technological Change,* Allen and Unwin, London.
67. See Mishan, E. J. (1967), *The Costs of Economic Growth,* Staples Press, London, Appendix A.
68. The traditional views about the benefits of improved technology and economic growth are well expressed in the conclusion of Stubbs, P. (1980), *Technology and Australia's Future: Industry and International Competitiveness,* AIDA Research Centre Publication, Melbourne. This monograph deals in some chapters with technology issues of policy importance throughout the world.
69. (a) Economic growth may be bringing the day of resource crisis and reckoning for mankind closer. See Daly, Herman (1979), Entropy, growth and the political economy of scarcity, in: *Scarcity and Growth Reconsidered* (ed. V. Kerry Smith), Resources for the Future, Baltimore, pp. 67–94.

(b) Mishan argues that continued technological growth and economic growth will not only wreck the ecological order of the world but also the *social order*. See Mishan, E. J. (1979), *The Economic Growth Debate,* George Allen & Unwin, London.

(c) See also Baumol, W. J. and Oates, W. E. (1979), *Economics, Environmental Policy and the Quality of Life,* Prentice-Hall, Englewood Cliffs, esp. Ch. 3 on global pollution thresholds and economic growth.

CHAPTER FOUR

Science and Technology Policy in Large OECD Economies

4.1 Introduction and background data

Government science and technology policies today have a wide social and economic impact. This impact is so great that no government can safely ignore the ramifications of its science and technology programmes nor avoid difficult choices for long. Electorates are increasingly realizing that almost every facet of modern life, the quality of our life, the future prospects for mankind and our hopes for a better world depend heavily on the nature of S & T change. Increasingly there is the fear that collective advances in S & T will enslave or master us rather than remain our servant. Can we be sure that new S & T advances will be employed to provide us with a better life however this may be defined? As discussed in earlier chapters we cannot be sure of this at all. The prospects of us becoming trapped by collective advances in S &T cannot be dismissed as a wild fantasy. This raises the question of alternative government policies.

The alternatives facing societies may not be all that palatable. At least three choices of government policies could be considered:

(1) Try to stop or slow down S & T change.
(2) Establish collective or social priorities for S & T change.
(3) Let S & T change proceed in a collectively unplanned, uncoordinated fashion in which the individual units within government and in the society at large such as companies pursue their own self-interest with scant regard to the common good.

Given the large social and economic overspills from new S & T, the last mentioned choice could result in our worst fears being realized. On the other hand to adopt the first alternative is to forgo the substantial advantages that can stem from new S & T. It would appear to be a regressive step. The second choice appears to be the most appealing but is not uncomplicated and without dangers. As considered earlier, how can collective priorities be determined? Would even limited collective optimization in this field of S & T planning exceed the rational capacity of man? Would it be better to blunder on as suggested by Lindblom? Who will have a say in determining the collective priorities to be espoused in government S & T policy and how will the opinions of different interest groups be taken into account? Will the priorities for S & T policy determined by government, possibly using participatory mechanisms, truly reflect the collective good however defined or will they reflect the self-interest of dominant power groups, indeed, strengthen the power of these groups to the possible detriment of society as a whole? How and to what extent is it desirable to encourage public participation in decision-making related to S & T as for example surveyed in *Technology on Trial*, OECD, Paris, 1979? Will increased government control of S & T restrict freedom and liberty and be worse than the technological restrictions on individual freedom that might be imposed by uncontrolled technological development? Societies are forced to choose between the above three policy choices, or variants on them, even if it is not always a conscious choice.

It is informative to consider the actual choices which governments have made in planning and formulating their S & T policies, how procedures and priorities compare between countries and how these have changed with the passage of time. Nine industrialized OECD countries have been selected for this purpose to provide a sample of government responses in industrialized non-communist countries to society's demands on government S & T policies. For comparative purposes, it is convenient to divide the selected countries into two groups for consideration. The policies of the four countries with large economies (West Germany, Japan, United Kingdom and United States) are discussed in this chapter and the policies of countries with smaller economies (Belgium, Canada, Netherlands, Sweden and Switzerland) are considered in the next chapter. The aim is to give a broad overview of the way in which government S & T policy is formulated, co-ordinated and administered in these countries and the priorities that governments have developed for such policies.

Before considering the S & T priorities of Germany, Japan, United Kingdom and the United States, it is worthwhile by way of background noting some of the broad comparative features of these economies and their science effort. The relative sizes of the economies, differences in the degree of their international dependence and the structure of their civilian employment may influence their scientific effort and science policies.

In terms of the size of its home market as indicated by the size of its GDP, the US economy is by far the largest. It is also largest in terms of population and area. The Japanese domestic market is the next largest in size, followed by Germany and the UK. However, the sizes of the German and UK markets

Table 4.1 Some indicators of the relative size of the economies of Germany, Japan, USA and UK – gross domestic product, population and area.

Country	GDP (Billion US $) 1979	Population* (thousands)	Area* (1000 km³)
Germany	755.8	61.3	248.6
Japan	1021.6	114.9	372.3
United Kingdom	391.2	55.9	244.0
United States	2349.0	218.5	9363.1

* In 1978.
Source Based on figures in *The OECD Observer*, March, 1980.

are much extended by their membership of the EEC. Table 4.1 sets out relevant figures indicating the relative sizes of these economies.

Germany, Japan and the UK exhibit much more external dependence than the USA. The share of their GDP accounted for by imports into the UK and Germany is comparatively large. Surprisingly enough, that for Japan is lower but the extreme shortage of agricultural land in Japan is one factor which makes Japan internationally dependent as does its shortage of energy resources. Both German and Japanese policy-makers have frequently commented on the international dependence of their economies, and this dependence has influenced their science priorities. Some measures of international dependence are given in Table 4.2 for Germany, Japan, UK and USA.

Table 4.2 Some possible indicators of international dependence – population density, population relative to agricultural area, imports as a percentage of GDP – Germany, Japan, USA and UK, 1978.

Country	Inhabitants per km²	Inhabitants per km² of agricultural land	Imports (goods only) as a percentage of GDP
Germany	247	404	18.9
Japan	309	1995	8.1
United Kingdom	229	301	25.4
United States	23	50	8.2

Source Based on figures in *The OECD Observer*, March, 1980.

Table 4.3 Structure of civilian employment by sectors – Germany, Japan, UK and USA, 1978.

Country	Agriculture, forestry and fishing %	Industry (mining and manufacturing)	Other (services) %
Germany	6.1	45.1	48.4
Japan	11.7	35.0	53.3
United Kingdom	2.7	39.7	57.6
United States	3.7	31.2	65.1

Source Based on figures in *The OECD Observer*, March, 1980.

Table 4.3 indicates that all the countries in this group are highly industrialized.

GERD as a percentage of GDP is roughly comparable for all of these countries, being around 2%. However, the absolute expenditure of the USA on R & D is by far the greatest of any of these countries being more than four times the level of Germany or Japan. The levels of German and Japanese expenditures on R & D are comparable, whereas that of the UK is lowest in this group. Absolute sizes of industrial R & D expenditure follow a similar comparative pattern. Details of these broad indicators of scientific effort are given in Table 4.4.

Slightly later figures for GERD, reported in *The OECD Observer* March

Table 4.4 Indicators of scientific effort – GERD as a percentage of GDP, absolute GERD, industrial R & D – Germany, Japan, UK and USA (1975 unless otherwise stated).

Country	GERD (% of GDP)	GERD (millions US $)	Industrial R & D as defined by OECD (millions US $)
Germany	2.1	8 857	5 881
Japan	1.7(76)	8 762	5 634
United Kingdom	2.1	(5 000)*	2 965
United States	2.3(76)	34 566	23 540

* Rough estimate.

Sources *The OECD Observer*, March, 1979; OECD (Winter 1977/78). *Science Resources Newsletter*.

1980, do not change the broad picture to any considerable extent. However GERD as a percentage of GDP rose in Japan in 1977/78 to 1.9% of GDP, the German figure was stable at 2.1% and the figure for the USA rose slightly in 1977 to 2.4%. Consequently, Japanese *absolute* expenditure on R & D moved ahead of German expenditure making Japan the third largest spender on R & D in the world after the USA and the USSR. In terms of absolute expenditure on R & D, the four countries under consideration are amongst the largest spenders in the world. The GERD of each of the smaller economies to be considered in the next chapter is less than one twentieth of that of the USA and less than a quarter of that of Germany and Japan.

With this broad background in mind, let us consider the science and technology policies and priorities of the Federal Republic of Germany, Japan, United Kingdom and the United States in turn.

FEDERAL REPUBLIC OF GERMANY

4.2 Articulation and administration of priorities in West Germany

Responsibility for research is divided in Germany between the Federal Government and the Länder (state governments) but financial considerations and the inter-regional impact of many research programmes have resulted in the relative importance of the Federal Government growing. The Länder under the Basic Law are responsible for education, the universities and libraries. In 1977, German governments contributed 13 250 million DM to support research and development. Of this, 7300 million DM (55%) was contributed by the Bund (Federal government) and 5950 million DM (45%) by the Länder. Most of the state contributions were for the support of research at universities and higher education institutions and for non-profit research institutes. Table 4.5 gives the contributions of German governments to R & D in Germany by sectors of performance for 1977.

The importance of the Federal Government in determining science and technology policy is possibly greater than its relative financial contribution indicates. Within the Federal Government the *Ministry of Research and Technology* (BMFT) plays a central role in science and technology policy. This Ministry (which is responsible for research policy) and the Ministry of Education and Science (which has some responsibility for university and education policy) were formed by Chancellor Schmidt in 1974. The BMFT has a co-ordinating role and an executive role in allocating and supervising many research grants.

The percentage of GDP devoted to GERD after rising strongly in Germany during the 1960s, failed to grow between 1971[1] and 1978. This percentage stood at just over 2% in 1978. Because of the relative freezing of

Table 4.5 Contributions of the German Federal and Länder Governments to R & D in Germany, 1977.

Sector performing R & D	Total government contribution (million DM)	(%)	Federal (million DM)	Länder (million DM)
Business enterprise	3 300	25	3 300	—
Higher education	5 150	39	520	4 630
Private non-profit research institutions	4 800	36	3 480	1 320
Total	13 250	100	7 300	5 950

Source (See ref. 7) Table 2, p. 7.

funds for R & D, there were a number of enquiries into S & T priorities in the early 1970s. Shortage of funds made it increasingly obvious that greater attention needed to be given to the optimal allocation of available funds for R & D. The three main reports emanating from these enquiries were:

(a) The *Grey Plan 1976–1978* issued by the German Research Society in 1976 and drawn up by scientists in part using a survey of researchers in universities.
(b) *Recommendations for the Organization, Planning and Promotion of Research* issued by the Science Council (an advisory body of federal and state government representatives and research association representatives). It argued that better co-ordination between the different research institutions is needed.
(c) *The Fifth Report of the Federal Government on Research* issued by the Federal Ministry for Research and Technology in 1976[2] states the aims of research policy as the government sees them and sets out government priorities in promoting research. The report provides a focal point for discussion of science and technology priorities in Germany.

Rainer Flohl claims that there is little parliamentary control over science policy, that there are few federal legislative hearings on science policy and that the legislature is lacking knowledge about science policy.[3]

On the other hand, the formulation of German science policy involves pluralistic elements and is not at the mercy of one Federal Ministry. The Länder have a voice and in part through them state organizations and private non-profit research institutions are strong and their controlling bodies represent a multiplicity of interests; other Federal Departments such as the Ministry of Economics have an interest, universities are a significant force and the interests of other bodies must be taken into account in formulating policy. The BMFT has a number of checks and balances even if these are not administered by the Bundestag. In the *Fifth Report*, the BMFT said:

> The Federation and the Länder, the German Research Society and the Foundations support research and development with a large number of promotion measures, thus safeguarding the plurality of research. Co-ordintion and co-operation in research are intended to secure the optimum use of resource, to establish appropriate priority areas of research and to help solve structural, organizational and financing problems.[4]

A special feature of German scientific effort is the existence of autonomous powerful science organizations. These include (a) the German Research Association (DFG), (b) the Max-Planck Society for the Promotion of Science (MPG), (c) the Fraunhofer Association and (d) various big science institutions. Non-profit research organizations of this type obtained more than 20% of government funds allocated to the support of R & D in 1976.

The DFG does not conduct research itself but provides research funds (grants) for German scientists for individual projects, for universities and research institutes. The MPG operates its own research institutes mostly in the area of basic science. The Fraunhofer Association is the executive organization for the Institutes of Applied Research. These 26 institutes employ about 2000 people and receive support from industry and government, and engage in contractual research.

Figure 4.1 indicates the flow of funds from government to research bodies. All of the bodies involved in these flows (payees and receivers) are in a position to influence science policy through formal and informal pressure groups.

In a recent report the OECD has said:

> Responsibility for R & D within the Federal Government is not lodged with one ministry alone. However, a co-ordination concept has been developed; each ministry has appointed a research co-ordination officer; they draw up statements of their activities, and they meet in an inter-ministerial committee under the chairmanship of the Federal Ministry for Research and Technology (BMFT). All their R & D projects are reported to the co-ordinating data bank DAVOR which is run by BMFT. Major research institutions supported by the Federal Government draw up programme budgets.[5]

Nevertheless, 83% of Federal funds for civil R & D are supplied by BMFT and it is responsible for research planning and co-ordination. BMFT supports a number of special programmes and at any one time may be funding more than 3800 projects. The cost of administering the funding of these is considerable. In 1976, BMFT's major outlays in decreasing size by

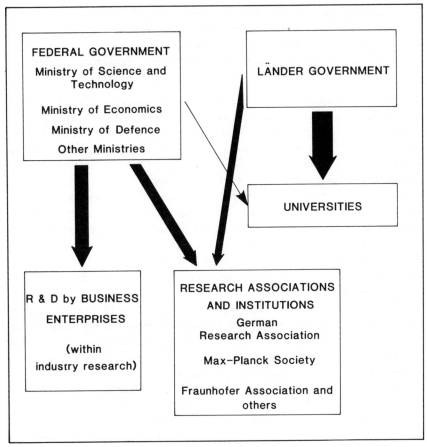

Fig. 4.1 Schematic representation of the flow of government funds for West German R & D. All bodies involved in this flow and others can influence R & D priorities.

programmes were for (1) nuclear programme, (2) space research, (3) data processing programmes, (3) programme for energy efficiency research and (5) transport and aviation research.

4.3 Selected features of German science and technology priorities

The clearest statement of the S & T priorities of the Federal German Government seem to be that in the *Fifth Research Report*[6] and exemplified in parts of *Faktenbericht 1977 Zum Bundesbericht Forschung*[7]. The prime aim of research activities and new technologies is seen by the German government to be the solving of societal problems. The German Federal Government believes that to do this, research should

(1) develop the efficiency and competitiveness of the national economy (so called modernization of industry goal);
(2) improve German living and working conditions;
(3) increase the efficiency of German science and research organizations;
(4) help maintain national security (contribute to defence).

The government feels that the first goal is necessary because 'due to world-wide economic trends and to the high degree of interdependence, the Federal Republic of Germany is confronted with the problem of securing and safeguarding the efficiency and competitiveness of its industry, and thus workplaces, by means of modernization.'[8] Economic competitive position and employment make this goal a first priority.

Considerable attention is to be given to improving the quality of life and environmental conditions under the second priority. 'Society and government are facing the task of improving living and working conditions and of eliminating the adverse effects of technological and social change'. Research is needed to deal with civilization-induced diseases and problems, improved working conditions, to provide better man-made environments and to preserve natural ones.

Basic research supports application orientated R & D and needs to be maintained at satisfactory levels if science efficiency is to be maintained. While not mentioned in the *Fifth Report* it is clear from *Faktenbericht* that defence is another important priority, even though it may be of diminishing importance as a research priority (see Table 4.6).

The relative expenditure of the Federal Government in support of these various goals and changes in planned expenditure gives an indication of the relative importance placed upon these priorities. Actual expenditures for 1975 and 1977 and planned expenditures for 1980 are shown in Table 4.6. This shows that the largest outlay of Federal R & D funds is for the modernization of industry (increasing its efficiency and competitiveness). This accounts for about 40% of Federal funds for R & D and is expected to rise slightly. Relative expenditure on living and working conditions has risen appreciably since 1972 and is expected to continue to rise with about 21% of Federal R & D funds being made available for this priority in 1980. Relative R & D expenditures (and real expenditure) for defence and security has declined and is expected to decline further. In 1972, 25% of Federal R & D funds were allocated to defence R & D; by 1977 this had fallen to 21.8% and by 1980 is expected to fall to 18.2%. Expenditure on the increase of scientific productivity (mostly basic research) rose slightly in relative terms between 1972 and 1977, and is expected to show an almost negligible decline by 1980.

Each one of these priority areas mentioned above is subdivided by fields of research expected to contribute to the main goal. The following sub-

Table 4.6 R & D outlays on priority areas by the German Federal Government 1972, 1977 and planned 1980 as a percentage of total R & D outlay by the Federal Government.

Priority Area	1972	1977	1980
1. Modernization of industry	42.0	39.2	41.4
2. Improvement of living and working conditions	14.7	19.4	21.2
3. Defence and security	25.0	21.8	18.2
4. Increased scientific productivity (mostly basic science)	18.0	19.0	18.6
5. Other	0.4	0.6	0.6
Total	100.0	100.0	100.0

Source (See ref. 7) Table 16, pp. 36, 37.

divisions have been made and their percentage of total Federal R & D outlay in 1977 is shown in brackets:[9]

R & D contributing to the modernization of industry
1. Securing of energy and raw materials (21.2)
2. Promotion of data processing (4.2)
3. Technical communication and electronics (1.9)
4. Innovative technologies and other key areas (4.7)
5. Space research and space technology (7.3)

R & D contributing to living and working conditions
1. Research on health and nutrition (7.0)
2. Humanization of work life and improvement of training facilities (1.7)
3. Improvements in the environment (5.3)
4. Improvement in transport and transport systems (5.3)

R & D contributing to increased scientific productivity
1. General promotion of research (basic research) (18.1)
2. Information and documentation (0.9)

The identification of key areas assists long-term planning and discussion of science policy. Of particular interest is the major emphasis of Germany on energy research. It has the second largest research effort in this area of any Western country. While its major emphasis is on research into nuclear energy, increasing attention is being given to other fields. A good deal of attention is being devoted to energy-efficiency research. This is concerned with heat–power relationships and making use of much heat (e.g. from power stations) which is now wasted. Research into coal utilization is also being expanded. In 1977, research designed to secure energy supplies and raw materials accounted for 21.2% of total Federal Government R & D spending and the German Government plans to increase this to 23.5% by 1980.

In the science and technology policy of the German Federal Government:

> the main emphasis is on the promotion of new technologies for future oriented branches of industry and the services sector, in order to safeguard the competitiveness of the economy. This fundamental objective is accompanied, as a complementary task, by programmes making increased allowance for social and societal problems in the solution of which research policy is expected to provide workable principles.[10]

The *Fifth Report* sets out the factors which the Federal Government takes into account in deciding to support R & D for the benefit of industry. Support is likely to be given[11]

1. when this R & D is necessary to facilitate the structural adjustment of industry in the light of changes in the international division of labour and the alteration of competition between countries;
2. when natural resources are developed or saved by the R & D, in particular energy and raw materials;
3. when the R & D provides large overspills to sectors beyond the performing sector;
4. when the technology from the R & D is likely to have favourable environmental effects, for instance reduce pollution or improve conditions in places of work;
5. when R & D is 'oriented towards an improvement of the fulfilment of public tasks and of the infrastructure, in particular areas such as public health, working environment, food supply and waste disposal, communications and transport'.

In particular government support for R & D is likely to be forthcoming, whenever

- the scientific and technical, as well as economic risks have to be rated as very high;
- large funds are required;
- developments will take a very long time, so that profits cannot be expected in the foreseeable future;
- the market underrates more advanced technological solutions since it responds not so much to the future as to the prevailing conditions as far as demand, shortage and constraints are concerned;
- demand does not suffice to bring about new technological solutions which are exclusively or predominantly in the interest of the general public or are applied in the sector.[12]

The German government provides assistance for industrial R & D in a diversity of ways[13] *some of which are not apparent from* its expenditures as, for example, set out in Table 4.5. It provides tax concessions to firms on their R & D expenditure, for example, and capital and credit assistance is available. The policies of the government are designed to promote:

(a) research, development and innovation within companies;
(b) research on behalf of or for the benefit of industry by research institutes and organizations, such as the Fraunhofer Association, under contract, or by research co-operatives;
(c) technology transfer.

Details of these schemes are set out by the Commission of the European Communities in *General Scheme of the Federal Republic's Research and Technology Policy for Small and Medium-sized Companies* 1978, and it is not possible to provide details here.

However, the methods of support for R & D within business enterprises can be briefly summarized. The main methods are:

(a) Direct promotion of projects
  (i) promotion of research and development projects under the specialized programmes of the Federal Ministry for Research and Technology (BMFT);
  (ii) the primary innovation programme to promote technologically novel and economically significant products and processes, which has been running since 1971, and the programme for technical development in Berlin industry – both under the Federal Ministry for the Economy (BMWi).

(b) Assistance with capital and credit
  (i) extensive risk-sharing through a private venture-capital company which provides equity capital and assistance to management in connexion with innovation in smaller companies;
  (ii) the provision of long-term, fixed-interest investment loans at low rates of interest to small and medium-sized companies under ERP (European Recovery Programme) schemes.

(c) Tax concessions
  Allowances for:
  (i) capital expenditure on research and development;
  (ii) bought-in intangible assets which have been capitalized (e.g. patents).[14]

Currently, a special effort is being made by the German government to assist small and medium-sized firms through research and technology policy. The general scheme evolved has only been decided on after much consultation.[15] New tax concessions for R & D will be relatively of greater benefit to smaller firms (the concessions are 'tapered') and bought-in capitalized intangible assets such as the right of use of patents will be allowed up to 500 000 DM annually as a tax deduction. Special funds are to be made available to the Fraunhofer Association for the following purposes:

(a) to finance part of the cost of R & D contracts for small and medium-sized firms which are not in a position to finance the whole cost unaided;
(b) to advise small and medium-sized firms about the application of new technologies and about the market prospects of new products;
(c) to contribute to the financing of, and to produce, studies and experts' reports on technical/economic problems.[16]

Although performance and funding of German S & T effort is pluralistic, progress has been made in rationalizing the system, co-ordinating and organizing it and BMFT has played a major role in this. Priorities have been specifically stated and have become a basis for planned and actual government action. Furthermore, German priorities have altered. As the OECD points out:

> On the whole, there has been a significant shift in orientation of S & T policy in comparison with the decade 1960–1970. The aims and priorities are now derived from a national goal-setting viewpoint, relying in the first place on the national potential for scientific discovery and technological innovation. In the past decade, many research objectives were dictated by political considerations or by international obligations such as civil space, civil nuclear or military R & D.[17]

Societal problems are now the pivot of German science and technology policy.[18]

JAPAN

4.4 Articulation and administration of priorities in Japan

Since the Meiji restoration, the government of Japan has recognized the great importance of S & T for economic development and has taken active measures to import technology and promote the development of new technology in Japan. Nevertheless, after World War II, Japan was economically and technologically behind advanced European nations and the USA. Its efforts to catch up were redoubled. National economic growth was given the first priority by Japan and all measures, including S & T technology measures, were focused on this goal.[19] The import of technology from abroad (rather than production of new domestic technology) and its assimilation in Japan was the major S & T plank in the economic growth programme.[20]

Industrial policy was closely tied in with the growth programme through MITI (Ministry of International Trade and Industry) and science and technological policies helped to support these industrial policies.[21] At first after the War, Japan concentrated on building up five basic industries (steel, shipbuilding, coal, power and fertilizers) and the government provided economic concessions for these industries. Partially as a result of the stimulus from the Korean War, the Japanese economy expanded remarkably during the 1950s and it was possible for the government to add more sophisticated strategic industries to its priority list. Electronics and automobiles were added in the 1950s and computers in the 1960s. During the 1960s Japan was also drawn into some of the big science fields such as nuclear energy and space research. However, broadly up to the 1970s the main emphasis of Japanese policy was on continuing, and high economic growth and S & T policy was designed to support this overriding aim.

But in the 1970s, Japan's circumstances changed. Japanese incomes were now high by world standards and policies to achieve the growth goal had been so successful that Japanese GDP became the second highest in the non-communist world. Furthermore, instead of suffering a chronic deficit in its balance of payments, Japan was faced by an 'embarrassing' surplus. At the same time, the Japanese people were demanding an improvement in the quality of life and environmental conditions and Japan (a country extremely deficient in energy resources) had a continuing energy crisis to cope with. A new direction in economic and social policy, and therefore S & T policy, seemed to be called for. Japan is under pressure to adjust its policies both from within (in favour of more quality of life considerations) and externally (principally from the USA) because of the impact of its international competitiveness.

In this respect Rapp has said:

> The response of Japanese policy-makers to these dual pressures, internal and external, has been to revise the hundred-year-old policy and to focus more on internal demands. Their plan is to reinflate the economy by increasing expenditures on long-needed social-overhead investments such as roads, improved harbours, parks and pollution control. In addition, imports will be liberalized, the rate of industrial-equipment investment will be reduced, export expansion will be de-emphasized, and overseas investment will be encouraged. The complexity of both the Japanese economy and the decision-making process has necessitated some lag between the perception of the need for change, policy formulation, and implementation. The first two steps now seem fairly complete, but policy implementation is only beginning and is encountering some difficulties.[22]

The changed circumstances of Japan also have other implications. Having caught up and in some cases surpassed other advanced industrial economies,

Japan can no longer rely on technology transfer as one of the main strings in its bow for economic development. The technology gap has largely disappeared, and the technologies now which Japan may seek to import are likely to be coveted by their recent developers overseas. In this regard, the Planning Bureau of the Science and Technology Agency has commented:

> While the international competitive power of Japan's industries has been steadily rising, the relative ease with which other nations provided technology has been gradually disappearing. In this regard, increasingly stringent conditions have been attached to the introduction of foreign technology; for example, in the conditions of royalty, and restrictions on export markets. Also, increasingly difficult demands may be attached as conditions, e.g. demands for cross licence agreements or for shares of company stock. [23] . . .
> [In short] it is anticipated that the introduction of technology from abroad, which hitherto has played such an important role in the growth of Japan's science and technology, will gradually become more difficult. Hereafter it is indispensable for Japan to promote independent technological developments.[24]

A number of government departments and bodies participate in the formulation of goals for government policy and for S & T policy. Three bodies, however, are believed to be most influential in the process of priority setting in S & T. These are: (1) the Ministry of Finance, (2) the Ministry of International Trade and Industry and (3) the Science and Technology Agency (STA) attached to the Prime Minister's Office. At the operational level STA and MITI are the most important bodies.

STA co-ordinates research activity *within ministries other than MITI*, takes part in determining their annual R & D budgets, operates six national laboratories, provides policy advice, holds funds for multi-ministerial projects and administers the Japan Research and Development Corporation (JRDC). Furthermore, it acts as secretariat for The Council for Science and Technology (CST), the Atomic Energy Commission, the Space Activities Commission and the Council for Ocean Development.

MITI has a wide mandate to plan and co-ordinate the effective use of national and foreign resources. As part of this task it is deeply involved in S & T policy. In this regard its main administering body is its Agency for Industrial Science and Technology (AIST) 'which oversees 16 research laboratories, administers national research programmes (NRDP) and is responsible for setting standards, while more recently it has developed a programme of technological evaluation'. MITI also regulates technology exports and imports, foreign investment, patenting and other activities. Its Agency for Small and Medium Enterprises provides small and medium firms with technical and financial assistance.

While STA co-ordinates the activities of ministries other than MITI, to some extent the other ministries carrying out research (Ministry of Education (basic research in universities), Ministry of Agriculture and Forestry, Ministry of Health and Welfare, etc.) have a considerable amount of independence in their R & D programmes.

The CST's role is to formulate long-term policy aims. In this role it helps to provide an overall approach to the S & T policies of STA and MITI and the other ministries. It makes recommendations direct to the Prime Minister.

The broad organization of Japanese government S & T is as shown in Figure 4.2. In this organization chart, the Environment Agency formed in 1971 is included. Its responsibilities include inter ministerial co-ordination of environmental protection technology, and it carries out a certain amount of environmental research.[25]

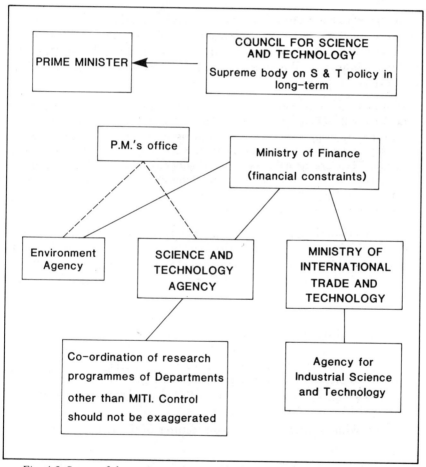

Fig. 4.2 Some of the main government bodies involved in Japanese science and technology policies and in the setting of priorities.

In the changed circumstances of Japan, the government is attempting to obtain a national consensus on new long-term goals for economic and social policy and S & T policy as a contributor to the achievement of national goals. Rapp observed in the early 1970s that there are:

> indications that a general policy decision is being reached via the well-known consensus process. That is, opinions are solicited from various elements of society and the economy (big business, small business, agriculture, labour, various ministries, politicians, and academics) until an overall agreement and policy emerge. At this point, although resistance and disagreements may continue, the policies will usually be pursued and implemented as each sector makes its contribution. This process facilitates consistency and accommodation in establishing Japan's structural goals and in carrying through programs.[26]

On the other hand, Andrea Boltho is cynical about the consensus process and very much doubts that it is capable of swinging Japanese social and economic policy around to significantly promoting the quality of life of the people and improving the environment. In his view bureaucracy, politicians and business in Japan form a grand coalition. Furthermore,

> ultimately, as in all capitalist societies, and Japan does not seem to be fundamentally different in this respect, the 'consensus' and the basic rules of the game are mapped out not so much by the democratic procedures as by pressures emanating from ruling interests, in this instance the business community. The working of the system has a logic of its own which it imposes on the political decision-making process. If anything, this influence may be more pervasive in Japan than elsewhere and there may be fewer countervailing powers – for complex historical and sociological reasons. . . .[27]

Nevertheless a new consensus had emerged by mid-1977. At that time, the CST was able to issue a report on the foundations of Japan's overall S & T policy based on long-term prospects entitled *Science and Technology Policy in the Age of Natural Resources Scarcity*. This report sets out the new targets of Japanese science and technology policy. The new direction of policies accord with those in many other economies.

> It is expected that Japan's future will have considerable difficulties in the problems of resources and environment, and it has become a great task assigned to science and technology policies to grasp social needs more exactly than heretofore, and to solve these problems by using the resources of science and technology. This is a movement not only in Japan, but in other countries as well.[28]

It is worthwhile considering the main features of the Report.

4.5 Selected features of Japanese science and technology priorities

The CST has put forward five main targets for Japanese S & T policy. These are:

1. to contribute to ensuring a stable supply of resources (energy, food, water resources and raw materials) and their economical use;
2. to help provide a desirable living environment and provide solutions to environmental and safety problems;
3. to contribute to the maintenance of health;
4. to produce pioneering and basic scientific techniques;
5. to foster techniques which provide a basis for Japanese international co-operation and/or secure the international competitiveness of Japanese industry.

These aims are not dissimilar to those of many other OECD countries and show considerable similarity with those of Germany for example.

The Council has outlined the main ways in which it sees the targets as being achieved or approached. Under target (1) for instance, energy is the Council's first concern and research programmes are under way or planned to ensure the better utilization of nuclear fuel, fossil fuels, solar energy, geothermal energy and hydrogen. The nuclear programme is principally under the control of STA and the non-nuclear under the control of MITI. Furthermore, generation and distribution, the use of waste heat, rationalization of the use of heat in industry and new forms of transport are to be considered. Similar details are given under target (1) for food supplies, water resources and raw materials and for the other targets.[29]

Under target (5) it is planned as far as international co-operation is concerned 'to offer science and technology to developing countries with special regard to their needs; and build up a mutual system of exchange of advanced science and technology with the advanced countries.'[30]

In order to secure the international competitiveness of Japanese industry it is planned to concentrate on the development of techniques and products, the output of which represents (a) a high value added in production, (b) the autonomous development of new products, techniques and systems and (c) techniques involving labour saving, rationalization and simplificiation. In more detail, the following are to be pursued:

> To attempt improvement in the techniques tending towards higher added value such as, electronic machinery, telecommunications, information treatment, aircraft, transport machinery and atomic power and aim at the improvement of their technology with regard to plants resulting from the combination of high degree and wide range of hard ware techniques and soft ware techniques.
>
> To strive for the development of techniques necessary for the development of original products by means of autonomous technical development and the development of new products resulting from the accumulation of elements,

multifunctioning and unification of machinery and electricity, and other systematizations.

To strive to discover and use techniques necessary to make the production process straight, capsulization of components, and the introduction of computers in order to encourage labor saving and rationalization throughout various industrial areas.[31]

Well thought out programmes also exist for the achievement of the other targets. The basic view underlying the new approach is that there should be greater *harmony* between society and the development and application of new technology, that policy must be *flexible* to cope with rapid change and unexpected socio-economic circumstances, the development and application of science and technology must be promoted on an *overall and systematic basis* and 'policy operation must be executed based on a *priority principle*'.

The CST maintains further, and this is in line with greater attention being paid to socio-economic objectives, that

In administering science and technology policies under these circumstances where science and technology become more closely relevant and involved with socio-economic situations, it is important to form a systematic policy-making attitude which takes into account elucidation of socio-economic phenomena, comprehension and analysis of people's view of life and values etc. In order to attain this purpose, it is necessary to positively incorporate the views and ideas of cultural and social sciences, and particularly to attach more importance to the promotion of soft science which can contribute to systematic policy – and decision-making.[32]

Other important recommendations on S & T priorities by the Council include:

(a) a recommendation that GERD as a percentage of GDP be raised in the short-term to 2.5% and in the long-term to 3%;

(b) that government spending on R & D increase to compensate for the fall-off in private expenditure by business as a result of the recession and to foster quality of life and environmental goals;

(c) that greater co-operation be encouraged between performers of R & D, that more R & D facilities be jointly used, that mobility of researchers be increased and that a group of 'freelance' researchers be established;

(d) large projects with a long gestation period and uncertain results should be undertaken as national projects;

(e) that greater attention be given to information collection and dissemination;

(f) that regional requirements continue to be taken into account in setting priorities and regional involvement be fostered;[33]

(g) that there be more evaluation of the use of and effects of techniques

and greater social feedback;
(h) that in accordance with the changed direction of science and technology policy that the composition of trained manpower and scientists be altered.[34]

Many different measures have been adopted by the Japanese government to encourage the fulfilment of its S & T technology priorities. These include taxation allowances and low cost loans to industry for R & D and technological innovation and replacement. Both specific and non-specific measures are taken. It is not possible to discuss these measures here.[35] Nevertheless, the NRDP, commenced in 1966 and expanded in scope, are worthy of special mention. These programmes have a high national priority.

AIST of MITI is responsible for running the programmes. Projects are carefully selected, their execution involves the intermeshing of performers in business, government and universities, and the prototype financed by the state is assigned to the commercial sector.

The projects are carefully selected according to the following criteria:

- the urgent nature of certain types of research aimed at upgrading the industrial structure, optimizing the use of natural resources, preventing industrial pollution, etc.,
- technologies which are expected to play a leading role (particularly relating to the manufacturing and industrial sectors),
- the risky and costly nature of certain R & D projects tending to discourage development by private industry,
- uncertainty as to the aims of certain programmes and the means required for achieving them;
- the need to meet growing social needs through co-operative action involving government, industry and the universities.

These projects are put out for tender and are mainly taken up by the large conglomerates. Negotiations take place not only between the AIST and the large groupings but also between the latter and the small sub-contractors and suppliers who will be involved in carrying out the future programme. The necessary work is commenced as soon as the official agreement has been signed and is normally carried out in AIST's 16 laboratories except for programmes in such fields as aeronautics or computers. Engineers and researchers from the private sector engaging in the project naturally take part in research activities and may use the facilities of the State institutes.[36]

This brief review indicates that Japanese S & T priorities have changed and that the Japanese government, subject to consensus, is altering S & T policy to reflect this. One of the basic principles of Japanese policy is that S & T policies must be operated in accordance with overall priority principles arrived at by consensus. Nevertheless, the needs of industry for S & T has remained uppermost in Japanese policy.

126 *Science and Technology Policy*

UNITED KINGDOM

4.6 **Articulation and administration of priorities in the United Kingdom**

The British Government has rejected the idea of overall, consistent national priorities for S & T policy and the establishment of a Minister or equivalent for R & D. As the OECD points out:

> In the United Kingdom objectives for science and technology policy are not centrally defined, as those for applied research and development are expected to flow from the research requirements associated with the particular responsibilities of individual ministerial departments. It is considered that priorities in fundamental research are best determined by the scientists themselves, with only the broadest strategic control policy formulation.[37]

Decisions about the volume of R & D spending are largely in the hands of individual ministers. As stated in Parliamentary Command Paper 5046:

> Decisions about the research and development required to support national economic and social policies must rest with Ministers who have responsibility for those policies. The Government's view, as stated in the White Paper on 'The Reorganisation of Central Government' (Cmnd. 4506), is that Government Departments should be organised by reference to the task to be done and the objectives to be met. Applied research and development are necessary to achieve many of the Government's objectives, but they cannot be regarded as forming a distinct function of government. Any attempt to formulate overall objectives for a supposedly collective activity of research and development would lead to confusion.[38]

Nevertheless, the Government feels that some interdepartmental co-ordination and co-operation is necessary in S & T. At first, after 1972 the Chief Scientific Adviser was given the role of overall review of R & D but this was later taken over by a committee of Chief Scientists and Permanent Secretaries of the Departments concerned under the chairmanship of the Deputy Secretary (Science and Technology) of the Cabinet Office. The Cabinet Office also has a Group which monitors the working of R & D arrangements, Parliament has a Select Committee on Science and Technology and the Central Policy Review Staff assists in suggesting some science priorities. The Lord Privy Seal through the Cabinet Office has an overall co-ordinating role. There is no attempt, however at close co-ordination and control of the R & D activities of the individual departments.

In order of their expenditure on R & D, the Departments with greatest outgoings are Defence, Industry, Education and Science and considerable amounts are spent by the Department of the Environment, Department of Energy, the Ministry of Agriculture, Fisheries and Food, and the Ministry of Overseas Development and other departments. Almost half of the Government's expenditure on R & D in 1975 was for defence purposes, a proportion similar to that of the USA and very high by world standards.

The science budget of the Department of Education and Science is allocated mainly to five research councils and to a lesser extent to the British Museum and the Royal Society. The University Grants Committee which helps support research in universities, also falls within the responsibility of this Department. This Committee along with the Research Councils provide a dual system of support to universities.

The science budget of the Department is usually allocated between the research councils according to the advice of the Advisory Board of the Research Councils (ABRC). This Board includes the Chairman or Secretary of each of the Research Councils, the Chairman of the University Grants Committee, the Head of the Central Policy Review Staff, together with representatives with a major interest in Research Council work and independent members drawn from the universities, industry and the Royal Society of London. (Councils also have *ad hoc* inter-Council co-ordinating committees). The five research councils are Agriculture (ARC) Medicine (MRC) Natural Environment (NERC) Science (SRC) and Social Science (SSRC). Allocations out of the science budget to the Councils and other bodies are shown in Table 4.7, but some councils such as the ARC receive a considerable amount of contract research from other Departments. All the Councils except SSRC have their own research establishments, all give research grants and contracts, a number support research institutes and all provide postgraduate awards. Their relative allocations for these purposes are shown in Table 4.8.

Table 4.7 Allocation of UK science budget of Department of Education and Science to research councils and related bodies, 1976/1977.

Council or body	£m	%
Agricultural Research (ARC)	18.33	8.5
Medical Research (MRC)	37.36	17.5
Natural Environment (NERC)	26.05	12.1
Science (SRC)	117.19	54.3
Social Science (SSRC)	11.18	5.2
British Museum	3.86	1.8
Royal Society	1.98	0.9
Total	215.95	100.0

Source (See ref. 40).

128 Science and Technology Policy

Table 4.8 Percentage allocation of budget funds of UK research councils for different purposes, 1976/77.

Purpose	% of budget funds of councils
Research grants and contracts	21.4
Research units	7.5
Research Council establishments	33.5
Research Council grant-aided institutes	5.4
Postgraduate awards	10.1
International subscriptions	15.5
Centrally supported schemes and administration	6.6
Total	100.0*

* May not add up to 100 because of rounding.
Source (See ref. 40).

Table 4.8 indicates that the major portion of science council funds goes to the Science Council and the major share of these funds is spent on big science (high energy physics, astronomy and space research). However, the share of the Science Council is being pruned back. There is disenchantment in the UK about the yields provided by big science in relation to the funds outlaid and a swing in priorities away from big science is under way. Just over half of the funds of the research councils are spent on their own research establishments and units and international subscriptions (principally by SRC) to bodies in which they participate.

> The SRC co-ordinates the British scientific space research programme, and provides the United Kingdom contribution to the European Organisation for Nuclear Research (CERN), the civil science programme of the North Atlantic Treaty Organisation (NATO) and part of the U.K. contribution to the European Space Agency (ESA).

However, it should not be forgotten that less than 15% of UK government funds for R & D are allocated to the research councils.

Figure 4.3 sets out some of the main bodies involved in government funding and organization of R & D in the UK. This Chart includes ACARD, the Advisory Council on Applied Research and Development, formed in 1976.

The Select Committee on Science and Technology has said of the Council:

> The Government has now gone some way towards correcting the excessive

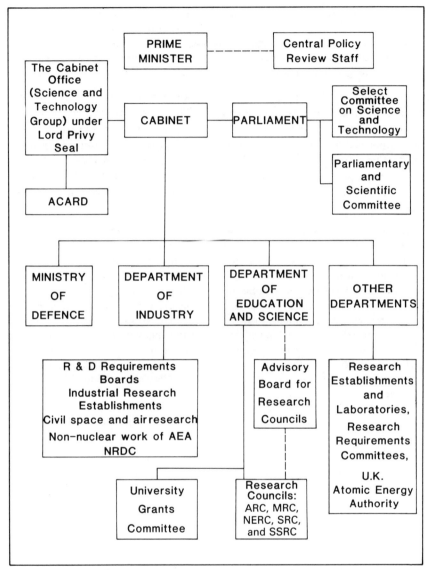

Fig. 4.3 Some of the main government bodies involved in the organization of R & D effort in the UK.[39]

emphasis on functional responsibilities by establishing an Advisory Council on Applied Research and Development, to be chaired by the Lord Privy Seal as Minister responsible for the co-ordination of government R & D. [One of its responsibilities] is the articulation of applied R & D with scientific research supported through the Department of Education and Science. We welcome this development and recommend that ACARD should review the relationship between government-supported applied R & D and government-funded basic research as a matter of urgency. In particular they should examine the operation of the customer–contractor relationship, and of the ABRC to ensure that effective machinery exists for relating basic science policies to long term departmental R & D strategies.[40]

Recent developments in British science policy can only be understood by reference to the Rothschild Report.[41] The Government adopted the customer/contractor principle for government research expenditure as recommended in this Report. The new framework was announced in July, 1972. The principle was aimed at strengthening the efficient *functional* operation of government and ensuring that R & D funded by departments was in accordance with their requirements. The Lord Privy Seal in his paper said:

> Ministers must ensure that departmental objectives are properly backed by applied research and development programmes, that is to say programmes directly related to these objectives. This country is fortunate in having a strong scientific and technical base. To make the best use of it means that those responsible for departmental objectives should also be responsible for defining their requirements in the clearest possible terms and commissioning the research and development work needed to achieve them. To do so the departmental "customers" must work in partnership with their research and development "contractors", whether inside or outside the Department. Responsibilities are then clear. Departments, as customers, define their requirements; contractors advise on the feasibility of meeting them and undertake the work; and the arrangements between them must be such as to ensure that the objectives remain attainable within reasonable cost. This is the customer/contractor approach. The Government reaffirms its intention, announced in the Green Paper, of extending it to all its applied research and development.[42]

Only applied research comes within the ambit of the Rothschild principle. Basic research is excluded. Customer Departments have a Chief Scientist (who generally acting on advice) advises the Minister and customers within his Department of research objectives. Customer departments have set up machinery (e.g. consultative organizations, research requirements committees or R & D Requirements Boards as in Department of Industry) to more closely define departmental needs in applied R & D. Contracts and commissions are defined in a considerable amount of detail and government research establishments, units and laboratories and other bodies may 'tender' for the work. 'The customer says what he wants; the contractor does it (if he can); and the customer pays'. As the British Council puts it:

> The 'customer–contractor' approach means that the Government department as 'customer' works in close collaboration with its R & D 'contractor',

Policy in Large Economies 131

whether the latter is inside or outside that department. 'Customers' define their requirements, contractors advise on the feasibility of meeting these requirements and carry out the work; collaboration between them ensures that the objectives remain attainable within reasonable cost. Individual Government departments have appointed Chief Scientists to co-ordinate their research needs. Government inter-disciplinary R & D is co-ordinated through meetings of departmental Chief Scientific Advisers chaired by the Deputy Secretary (Science and Technology) in the Cabinet Office.[43]

Reactions to the application of the Rothschild customer–contractor principle have been varied. Some government research bodies (NERC) find that the application of the principle has made it more risky for them to embark on long-term research projects, especially when these projects are being funded from several sources. Customers appear to be satisfied by the principle, but some of the contractors are worried about its operation. It has certainly created much more paper work. Furthermore, the Select Committee on Science and Technology remains concerned. It reported in 1976:

> In 1972 our concern was to ensure that government R & D strategy was not distorted by short-term conceptions of need flowing from the exercise by departments of their proxy 'customer' role, and that long-term needs, not particularly related to the needs of individual departments, were not thereby ignored. The creation of the new Committee of Chief Scientists and Permanent Secretaries, and of ACARD, indicates that the Government now recognise this danger and are strengthening the machinery for examining longer-term research and manpower requirements at an interdepartmental and extra-departmental level. We remain concerned about the need to ensure that adequate political control is exercised over R & D decisions which may have profound long-term effects on the community, but we believe that only time will demonstrate whether the new co-ordinating machinery is adequate for this task.[44]

The setting of priorities for British S & T remains a contentious issue. If the *Third Report from the Select Committee on Science and Technology*[45] is an indication, pressures exist in Parliament for improvements in the setting of priorities for the British government S & T effort. Pressures also exist for greater collaboration between universities and industry and for more emphasis on engineering and applied science, rather than basic science.

4.7 Selected features of United Kingdom science and technology priorities

The claim is not infrequently made in Britain that Britain gives insufficient attention to applied R & D. This was a concern in the Rothschild Report, is repeated in the reports of the Select Committee on Science and Technology, and was expressed by Sir Iewan Maddock, former Chief Scientist UK

Department of Industry on his retirement in 1977. He is reported to have said:

> The mistakes of the past were to place too much emphasis and to commit too much of the available money to too many big projects, aircraft, rockets, etc. Many of these projects were subsequently abandoned because the country's resources were overstretched. Not only did these projects extend the country financially but they concentrated too much of the scientific talent available in too few projects. The argument is that more of Britain's top scientists should be applying themselves to the real problems of the country rather than to fields most exciting to the scientists themselves. The question is, how to encourage scientists to do this when the prestige and professional honours are mostly in research removed from immediate practical relevance.

However, in comparison to Germany and Japan the *actual* composition of the British R & D effort indicates that it is strongly orientated towards application. A much lower proportion of research is carried out in universities and research institutes in Britain than in either Germany or Japan. Furthermore, in 1975 the only OECD country to surpass the UK in terms of the percentage of its total industrial resources allocated to R & D was the USA. The UK industrial research intensity stood at 1.75%, marginally in excess of that for Sweden and Switzerland, ahead of Germany and considerably in excess of Japan.[46]

A breakdown of public (government) funding of R & D by socio-economic purpose indicates that there is much less emphasis on the advancement of knowledge (basic research) in the UK than in Germany, Japan, Belgium, Sweden and most other advanced economies. Table 4.9 provides a comparison with Germany and Japan and indicates that the proportion of public funds for R & D provided for industrial growth is much greater in the UK than either in Germany or Japan. However, in view of the very high proportion of public R & D funds allocated to defence in the UK the absolute amount of German public funds for R & D for industrial growth is higher than that for the UK. On the other hand, if attention is purely concentrated on the allocation of public funds in support of civil R & D, the UK relative weight in favour of industry is markedly greater than in either Germany or Japan and weighted against the advancement of knowledge (basic research). Can we deduce from this that the functional system of allocation of public funds for R & D in the UK has led to excessive emphasis on short-term applied problems? Are dangers involved in Departments looking only to their own most pressing needs or political dictates in giving contracts for R & D? These are not easy questions to answer.

The UK has been very innovative in the measures adopted by it to stimulate research and innovation in industry. It is not possible to discuss these measures here but they include grants to industry research associations

Table 4.9 Public R & D funding in 1975 by socio-economic objectives – percentage allocation of funds.

Socio-Economic Objective	United Kingdom	Germany	Japan
Agriculture, forestry, fishing	4.4	1.9	13.5
Industrial growth	12.4	7.4	6.2
Production of energy	7.3	10.6	8.5
Transport and telecommunications	0.8	1.5	1.5
Urban and rural planning	1.7	1.1	1.1
Environment protection	0.5	1.0	1.4
Health	2.7	3.3	3.1
Social development services	1.0	4.9	1.2
Earth and atmosphere	0.7	1.8	0.7
Advancement of knowledge*	19.9	51.0	55.3
Civil space	2.3	4.3	5.3
Defence	46.4	11.1	2.3
Total†	100.0	100.0	100.0

* Includes general university funds.
† May not add up to 100% because of rounding.
Source Based on the table on p. 11 OECD (Spring 1977), *Science Resources Newsletter*.

and loans to firms for modernization of equipment and risk-sharing in new developments by firms such as in the development of new aircraft engines.[47]

The National Research and Development Corporation (NRDC) established by the Government in 1948 has had a useful impact on innovation in industry and on the development and exploitation of inventions. This government corporation is in certain matters responsible to the Minister of State for Industry. It was the model for similar corporations in other countries such as Japan and Canada. The Corporation provides finance for innovation in industry. The Annual Report of NRDC for 1976–77 points out that:

> The Corporation's financial support for innovation in industry normally takes the form of a joint venture in which the ownership of the invention and the responsibility for its development and subsequent exploitation remain with the company concerned. In a typical joint venture, the Corporation contributes an agreed proportion of the development expenditure in exchange for a levy on sales of the resulting product or some other form of return which reflects the risk-bearing nature of the support. The expenditure to which the Corporation contributes may include the costs of launching a new product on

the market and the associated working capital requirements. In these cases the Corporation may support a fixed proportion of the maximum negative cash flow of the project as a whole.[48]

In certain cases where a new company is set up to exploit an invention, NRDC may take a share of its equity capital.

The NRDC also has a number of inventions assigned to it. Most of these are from public sources. NRDC may patent these and attempts to encourage their commercial use. In many cases further development may be needed prior to commercialization and NRDC may provide funds for this purpose. Most of its assigned inventions come from public research establishments, research councils, universities and a number of other public bodies. A few (very few) come from private inventors.

In 1977, after tax and interest, NRDC earned a profit of £6.4m and it now has a prospect of achieving a surplus on its operation since inception. The major source of income for NRDC is from licences and its second most important source is from levies from joint venture projects.

The UK's approach to S & T priorities is pluralistic. Whether or not this is a 'luxury' for other than a superpower such as the USA is difficult to decide. As mentioned in Chapter 1 of this monograph, the comprehensive approach to policy-making also has its pitfalls, and some compromise between the extreme approaches may be called for. Britain appears to be searching for a suitable compromise.

UNITED STATES OF AMERICA

4.8 Articulation and administration of priorities in the United States

In the US national paper prepared for the 1979 UN Conference on Science and Technology for Development, the philosophy behind science and technology policy and its pluralistic nature in the USA is expressed as follows:

> The ways in which science and technology have evolved in the United States reflect the philosophical environment in which they have emerged – one that encourages freedom of thought and expression, a diversity of sources of competition and innovation, and the constructive development of individual talent and institutional resources.
>
> Throughout U.S. history, there has been no long-term, overarching design or plan for scientific and technological development. Progress was incremental, a response to specific situations – from the early challenges of westward expansion, the mapping of the country's geography, and inventorying its natural resources to the more recent adventures of space exploration. Such challenges, plus those of military defense and the pursuit of

economic development, have provided powerful motivation for the scientific community. And as economic growth has progressed, it has been possible to devote increased resources to science and technology.

A multiplicity of institutions – private and governmental – has emerged. This institutional pluralism has proved to be one of the major strengths of the United States in science and technology. On the one hand, these institutions have stimulated each other through competition; on the other hand, their respective strengths have had a mutually reinforcing effect.[49]

Nevertheless, the USA is moving towards improved co-ordination and more specific setting of goals for public support of its science and technology effort. In this respect, the *National Science and Technology Policy, Organization, and Priorities Act of 1976* is of considerable significance. It sets out broadly the goals which S & T should contribute to. In the Act, Congress sets out principles, methods of implementation and procedures to be followed in US S & T policy, establishes or provides for the establishment of the Office of Science and Technology Policy, the President's Committee on Science and Technology, and the Federal Co-ordinating Council for Science, Engineering and Technology. It also provides for an annual Science and Technology Report and for the Office of Science and Technology Policy to provide annually a five-year outlook for S & T in the United States. These measures should ensure that there is greater co-ordination of S & T policy than previously and greater co-ordination than exists in the UK for instance.

The first three clauses of Section 101(b) of the *National Science and Technology Policy, Organization, and Priorities Act of 1976* declares:

(1) The Federal Government should maintain central policy planning elements in the executive branch which assist Federal agencies in (A) identifying public problems and objectives, (B) mobilizing scientific and technological resources for essential national programs, (C) securing appropriate funding for programs so identified, (D) anticipating future concerns to which science and technology can contribute and devising strategies for the conducting of science and technology for such purposes, (E) reviewing systematically Federal science policy and programs and recommending legislative amendment thereof when needed. Such elements should include an advisory mechanism within the Executive Office of the President so that the Chief Executive may have available independent, expert judgment and assistance on policy matters which require accurate assessments of the complex scientific and technological features involved.

(2) It is a responsibility of the Federal Government to promote prompt, effective, reliable, and systematic transfer of scientific and technological information by such appropriate methods as programs conducted by non-governmental organizations, including industrial groups and technical societies. In particular, it is recognized as a responsibility of the Federal Government not only to co-ordinate and unify its own science and tech-

nology information systems, but to facilitate the close coupling of institutional scientific research with commercial application of the useful findings of science.

(3) It is further an appropriate Federal function to support scientific and technological efforts which are expected to provide results beneficial to the public but which the private sector may be unwilling or unable to support.[50]

Clearly one of the main purposes of the Act is to provide for a more comprehensive and more central approach to S & T policy and to improve the co-ordination and direction of public effort in S & T.

The Act also set out procedures which are likely to assist in the implementation of policy. These include (1) the use of Federal procurement policy, (2) the employment of explicit criteria, including cost-benefit principles where practicable, to identify the kinds of applied research and technology programmes that are appropriate for Federal funding and determine the extent of such support, (3) improved Federal patent policies and (4) indicates that the Government while maximizing the beneficial consequences of technology should act to minimize foreseeable injurious consequences.

The Act establishes the Office of Science and Technology Policy (OSTP) within the Executive Office of the President. The Director of the Office will advise the President on S & T policy and Federal science budgets, 'assist the Office of Management and Budget with an annual review and analysis of funding proposed for research and development in budgets of all Federal agencies, and aid the Office of Management and Budget and the agencies throughout the budget developing process', and assist the President in co-ordinating the R & D programmes of the Federal Government. The Director of OSTP has a wide brief on S & T policy planning, analysis and advice. These include, amongst several other tasks specifically set out in Sec. 205 of the Act, 'to seek to define coherent approaches for applying science and technology to critical and emerging national and international problems and promote co-ordination of the scientific and technological responsibilities of the Federal departments and agencies in the resolution of such problems'.

The 1976 Act also provides for the appointment of a President's Committee on Science and Technology to review S & T policy in the USA. The Committee is to give a final report within two years from its formation on aspects of Federal S & T policy and is comprised of a wide range of interest groups. The tasks of this Committee are extremely germane to the subject matter of this present monograph and it is worthwhile therefore stating these tasks in full. Section 303(a) of the US Science and Technology Act 1976 states:

> The Committee shall survey, examine, and analyze the overall context of the Federal science, engineering, and technology effort including missions,

goals, personnel, funding, organization, facilities, and activities in general, taking adequate account of the interests of individuals and groups that may be affected by Federal scientific, engineering and technical programs, including as appropriate, consultation with such individuals and groups. In carrying out its functions under this section, the Committee shall, among other things, consider needs for –

(1) organizational reform, including institutional realignment designed to place Federal agencies whose missions are primarily or solely devoted to scientific and technological research and development, and those agencies primarily or solely concerned with fuels, energy, and materials, within a single cabinet-level department;

(2) improvements in existing systems for handling scientific and technical information on a Government-wide basis, including consideration of the appropriate role to be played by the private sector in the dissemination of such information;

(3) improved technology assessment in the executive branch of the Federal Government;

(4) improved methods for effecting technology innovation, transfer, and use;

(5) stimulating more effective Federal–State and Federal–industry liaison and cooperation in science and technology, including the formation of Federal–State mechanisms for the mutual pursuit of this goal;

(6) reduction and simplification of Federal regulations and administrative practices and procedures which may have the effect of retarding technological innovation or opportunities for its utilization;

(7) a broader base for support of basic research;

(8) ways of strengthening the Nation's academic institutions' capabilities for research and education in science and technology;

(9) ways and means of effectively integrating scientific and technological factors into our national and international policies;

(10) maintenance of adequate scientific and technological manpower with regard to both quality and quantity;

(11) improved systems for planning and analysis of the Federal science and technology programs; and

(12) long-range study, analysis, and planning in regard to the application of science and technology to major national problems or concerns.[51]

The new measures and investigations under way in the USA are designed to improve the management of the public S & T effort. The provision of adequate overall information on S & T and policies associated with it is important for improved management. The provision in the 1976 Act for an Annual Science and Technology Report and an annual Five Year Outlook for S & T and associated policies should assist considerably.

The US Science and Technology Act 1976 indicated that these reports should be prepared by the Director of the OSTP. In 1977 this responsibility was transferred to the National Science Foundation (NSF). Nonetheless OSTP played a major part along with NSF in preparing the first Annual

Report on Science and Technology,[52] which President Carter submitted to Congress in September, 1978.

Chapter 1 of this Report was prepared by OSTP and points to a number of policy issues of significance in the US. These points include:

(a) We need a better definition of, and greater consensus on, our long-term goals in R & D. . . . Although R & D should neither be disengaged from the Federal budget cycle nor be locked into inflexible plans, the design and emphasis of our R & D programs should not be rehashed and redirected in each budget cycle. A longer-term approach would allow sufficient continuity so that programs could build momentum.

(b) Many believe that by providing incentives, or reducing risks and uncertainties, industry could and should be brought into the development phase much earlier. It is argued that industry can be far more effective in development than government because of its greater sensitivity to market forces.

(c) Another strategic issue that warrants continued examination is enhancing the linkage between the performers of research, in particular between industry and universities . . ., it is important that industry – university relationships be strengthened and that technology transfer mechanisms be improved.[53]

(d) There is to be a cabinet-level study of innovation in USA.

A rather unusual body, in that it is responsible directly to Congress, in US S & T policy, is the Office of Technology Assessment (OTA). It was created in 1972 to report to Congress at an early stage on the impacts of new technologies on society. It is to take account of beneficial and adverse impacts (physical, biological, economic, social and political) of technology and is to 'bring a long-term global and comprehensive perspective to bear and to provide Congress with independent, authoritative, even-headed assessments'. The OTA is responsible to a Congressional Board (with equal representation from the House of Representatives and the Senate) of which Senator Edward Kennedy is at present chairman. The present Board is reported to have given a new fillip to OTA.

Each year, OTA selects a number of technologies for in-depth study and assessment of their effects on society. While originally these technologies were nearly all proposed by the Board, a new process of selecting projects for assessment was instituted in 1978. A wide range of people were canvassed to suggest critical technological issues for study and report.

> From efforts to reach as broad and informed public as possible, OTA received 1,530 suggested topics for study. Another 2,875 items were extracted from the published literature. To cope with this large list, OTA mobilized its staff to organize, combine, winnow and rank the candidates into a manageable list of 30 items.

For each of the 30 items selected for assessment,[54] OTA gives a brief outline of the problem and a description of the method of assessment intended. The assessment is reported to Congress for possible action. The assessments to be made (in order of priority) are shown in Table 4.10. It is difficult to know at this stage how influential OTA and Congress will be in matters of technology application.

The Federal Departments and agencies in the USA with the largest outlays for R & D are, in order of outlay, the Department of Defence (DOD), the National Aeronautics and Space Administration (NASA), the Department of Energy (DOE), the Department of Health, Education, and

Table 4.10 Projects for assessment in order of priority to be undertaken by the Office of Technology Assessment in 1979.

Priority	Project	Priority	Project
1	Impact of Technology on National Water Supply and Demand	15	Allocating the Electromagnetic Spectrum Globally
2	Alternative Global Food Futures	16	Implications of Increased Longevity
3	Health Promotion and Disease Prevention	17	Controlled Thermonuclear Fusion
4	Technology and World Population	18	Technology and Mental Health
		19	Technology and Education
5	Impact of Technology on Productivity of the Land	20	Prescription Drug Use
		21	Forest Resource Technologies
6	Impacts of Technology on Productivity, Inflation and Employment	22	Health Technologies and Third-World Diseases
7	Technology and the Developing World – Meeting Basic Needs	23	Electric Vehicles: Applications and Impacts
8	Peace Technology	24	R & D Priorities for US Food Production
9	Impact of Microprocessing on Society	25	Alternative Materials Technologies
10	Application of Technology in Space	26	Deep Ocean Minerals Development
11	Designing for Conservation of Materials	27	Energy Efficiency in Industry
12	Future of Military Equipment	28	Role of Technology in Meeting Housing Needs
13	Impact of Technology on the Movement of Goods	29	Ocean Waste Disposal
14	Weather and Climate Technology	30	Technology and the Handicapped

Source (See ref. 54) Page v.

Welfare (HEW), the National Science Foundation (NSF) and the Department of Agriculture (USDA). Taking these departments and agencies into account, the main public bodies involved in US S & T policy are as indicated in Fig. 4.4. However, there are also committees of the Congress which regularly conduct hearings and issue reports on S & T. For instance, the Committee on Science and Technology, US House of Representatives, is particularly active in this regard and has the task of

> reviewing and studying, on a continuing basis, all laws, programs, and Government activities dealing with or involving non-military research and development. This Special Oversight function is to be performed in addition to the legislative and direct oversight function of the standing committees.[55,56]

4.9 Selected features of American science and technology priorities

GERD as a percentage of GDP was stable in the USA at 2.3% between 1973 and 1977 having declined from its peak of 3% in 1964. The US government does not wish this percentage to fall lower.

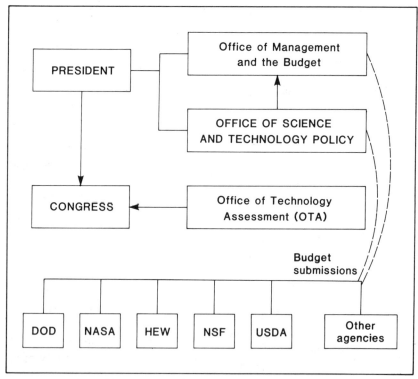

Fig. 4.4 Some of the main public bodies involved in science and technology policy in the USA.

During the last decade or so the USA has continued to show a high priority for security and defence and a little over half of the R & D funds of the Federal Government has been used for defence. Between 1968 and 1979 the proportion of Federal Government R & D funds for defence fell slightly from 52% to 50% of total Federal Government funding of R & D. The proportion used for civilian R & D rose markedly from 21% in 1968 to 39% in 1979 whereas the emphasis on space research plumeted from 27% of total Federal Government funding of R & D in 1968 to 11% in 1979.[57] These broad trends are indicated in Fig. 4.5 and indicate revealed priorities. The increased emphasis on civilian R & D is consistent with the pattern in most OECD countries.

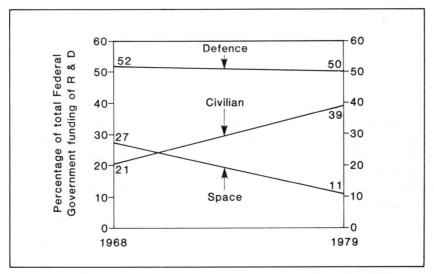

Fig. 4.5 A comparison of relative US Federal Government Funding for R & D for defence, civilian and space purposes, 1968 and 1979.

Over the last four decades, the US Federal Government's funding has revealed a reduced priority for funding development and an increased priority for funding basic and applied research, as indicated in Table 4.11. Although there was a slight drop in relative emphasis on basic research between 1970 and 1973, its relative weighting has been regained.

In the 1970s, the American Federal Government's expenditure on R & D in energy development and conversion grew at a most rapid rate; R & D dealing with health also rose substantially as did that dealing with the environment. These were all areas of greater priority in the 1970s than in the 1960s. The budget authority for 1979 (Table 4.12) gives some indication of current priorities by function in Federal funding of R & D. The current priority order is: (1) national defence, (2) space research and technology,

Table 4.11 Percentage allocation of Federal Government R & D funds between basic research, applied research and development in selected years.

Year	Basic research	Applied research	Development
1960	8	19	73
1968	16	19	65
1973	16	21	62
1978 (est)	17	23	60

Source (See ref. 52) Table 6.3, p. 44.

(3) energy, (4) health, (5) general science and basic research, (6) natural resources and the environment, (7) transportation, (8) agriculture and others as indicated.

However, it would be a fallacy to believe that government expenditures fully reflect government priorities. As was pointed out in earlier chapters, there are a number of other means by which a government can affect the S & T effort of a country. Tax concessions to businesses are one means. In the USA R & D expenditures may be deducted from taxable income in the year in which they are incurred and pollution control equipment can be amortized within sixty months. The Small Business Agency is also able to provide capital by way of equity or loans to small companies for investment in innovation. There are in addition many other measures to stimulate innovation in industry but these cannot be covered here.[58] But it should be noted that the American government has a strong preference in favour of contracting out its R & D to industry and this has provided a major benefit to American industry.

The United States government has, however, not failed to give growing emphasis to research in universities and colleges and Federally funded research and development centres (FFRDCs) associated with universities. The average growth rate in funds for these bodies from Federal sources for both basic and applied research have been greater than for any other performing sector (government or industry).[59] As mentioned earlier, the American Federal Government is concerned to improve links between industry and the universities and improve technology transfer mechanisms between the two. Nevertheless, the universities and colleges and associated FFRDCs have a high priority as R & D performers in the American system, more so than in the UK.

With the passing of the US Science and Technology Act in 1976 and with other measures the USA has moved towards greater comprehensiveness in its setting of S & T priorities. Nevertheless, without first hand experience it

Table 4.12 Budget authority for research and development 1979 (by budget function) of US Government.*

Function	1979 ($ millions estimate)
National defence	13 803
Space research and technology	3 751
Energy	3 186
Health	3 015
General science and basic research	1 135
Natural resources and environment	885
Transportation	799
Agriculture	511
Education, training, employment, and social services	279
Community and regional development	126
Veterans benefits and services	113
Commerce and housing credit	82
International affairs	79
Administration of justice	47
Income security	42
General government	9
Total	27 862

* Excludes R & D plant.
Source (See ref. 52) Table 3.1, p. 11.

is difficult to gauge the extent to which this process has progressed. Indeed, a recent US National Paper has stated that:

> The Federal Government does not have a comprehensive research and development budget or a comprehensive science and technology policy. Funding is provided in the budgets of 29 separate governmental departments and agencies; 12 have annual research and development budgets of over $100 million each. The White House Office of Science and Technology Policy maintains an overview of major programs and assists in setting priorities.[60]

4.10 Some observations

Ronayne has pointed out that in the USA:

no single department or agency is responsible for the funding or performance of governmental R and D; many departments are responsible for funding, contracting out and performing the research and development that they need in order to fulfil their assigned missions; there is no control controlling Department of Science Policy or Science, no interdepartmental co-ordinating committee and no official external advisory committee.[61]

While US science policy still remains pluralistic and decentralized some co-ordination is being attempted through the Office of Management and the Budget, and the OSTP. Both of these bodies are within the Executive Office of the President. OSTP and the President have emphasized the 'need for better definition of, and a greater consensus on long-term goals in R & D' in the USA. The US trend is towards more information on *overall* government science policy and science expenditures and indicators, greater articulation of national priorities and increased co-ordination in science funding. But the speed of change could be slow given the nature of the US system.

These trends are more strongly present in Germany and have been features of Japanese science policy for a considerable time. Both countries seem to have gone further than the USA in articulating national priorities and co-ordinating their scientific effort by a combination of consensus, concerted action and centralized control over large or vital parts of public R & D spending. Japanese science policy has been and is closely linked with its industrial policy. Its industrial policy is carefully formulated and targets are clearly defined. Largely the aim is to satisfy Japan's perceived need for its industry to be competitive and remain competitive internationally. Similarly German science policy gives a high priority to the 'the maintenance and competitiveness of the national economy'. But of course, and as pointed out above, the international competitiveness of domestic industries is not the only objective in the formulation of the government science policies in these countries. Nevertheless, it is hard to avoid the impression that it is the dominant objective.

While Britain at one time accepted the desirability of a fairly centralized system of government funding and control of science policy (based upon the Haldane Principle),[62] in 1972 it rejected the *desirability* of centralized co-ordinated control of government science effort and policy in favour of leaving science policies to individual government ministries and agencies in pursuit of their own goals. Departments were encouraged to specify their goals more closely using the customer–contractor principle. On the whole the British trend (except in the demand for priorities to be more explicit) is quite different, as we shall see, from the trend in the other countries in this sample.[63] Because of this approach no government science budget is available in the UK and statistics and data are patchy on the *overall* UK contribution to scientific effort. Presumably little value is seen in collecting overall data since it is not a basis for policy decisions.

Both the USA and the UK have become increasingly concerned about the international competitiveness of their industries but science policy approaches to dealing with this have been piecemeal. It might also be noted that the proportion of GERD spent on defence is much higher in the USA and UK than in Germany or Japan. This may place a relative burden on the USA and UK as far as their international industrial competitiveness is concerned despite the possible spin-offs to industry of such research and despite the fact that some export sales of defence equipment, for example fighter aircraft, may be obtained.

While it is difficult to make international comparisons, it would seem that the S & T policy of the USA is more comprehensive than that of the UK but less comprehensive than the S & T policies of Germany and Japan. All of these countries are and have been concerned in the current economic and social circumstances about S & T priorities and assessment. While there are some differences in the problems which they have had to face, one is struck by the extent to which their problems are similar. Germany and Japan appear to have gone further than the USA and the UK in explicitly mapping out their S & T priorities. We shall be able to assess the science policies of these four large economies further once we have considered the policies of the five countries with smaller economies.

Notes and references

Federal Republic of Germany

1. For a review of German science policy in the 1960s see Keck, O. (1976), West German science policy since the early 1960s: trends and objectives, *Research Policy*, 5, 116–17.
2. Ministry for Research and Technology (1976), *Fifth Report of the Federal Government on Research*, Bonn.
3. Flohl, R. (1977), The Federal Republic of Germany, *Science and Government Report International Almanac 1977* (ed. D. S. Greenberg), Science and Government Report Inc; Washington, pp. 51–58, especially p. 53.
4. *Fifth Report* (see ref. 2) p. 16.
5. OECD (1978), *Science and Technology Policy Outlook*, Paris, p. 55.
6. *Fifth Report* (see ref. 2).
7. Federal Minister of Science and Technology (1977), *Faktenbericht 1977 Zum Bundesbericht Forschung*, Bonn.
8. *Fifth Report* (see ref. 2) p. 5.
9. *Fakrenbericht* (see ref. 7) Table 16, pp. 36–37.
10. OECD (see ref. 5) p. 19.
11. *Fifth Report* (see ref. 2) p. 8.
12. *Fifth Report* (see ref. 2) pp. 8, 9. A number of the general arguments for such

assistance are covered in Chapters 1–3 of this monograph.
13. For general discussion of this subject see Oppenlander, K. (1977), The role of business and government in the promotion of innovation and the transfer of technology, in *Industrial Policies and Technology Transfers Between East and West* (ed. T. Saunders), Springer-Verlag, Vienna.
14. Commission of the European Communities (1978), *General Scheme of the Federal Republic's Research and Technology Policy for Small and Medium-Sized Companies*, pp. 18, 19.
15. Commission of the European Communities (see ref. 14) p. 5.
16. (a) Commission of the European Communities (see ref. 14) p. 32.
(b) For an earlier study of policies to promote industrial innovation in Germany see OECD (1978), *Policies for the Stimulation of Industrial Innovation*, Vol. II-1, Paris.
17. OECD (see ref. 5) p. 19.
18. For a recent overview of research and scientific organizations in Germany see Hauff, Volker and Haunschild, Hans-Hilger (1978), *Forschung in der Bundesrepublik Deutschland*, Kohlhammer, Stuttgart.

Japan

19. For a general review of science and Japanese government policy see *Outline of the White Paper on Science and Technology: Aimed at Making Technological Innovations in Social Development* (1977), Science and Technology Agency, Japan (unofficial translation by Foreign Press Centre), Ch. 1.
20. See, for instance:
(a) Ozawa, T. (1974), *Japan's Technological Challenge to the West, 1950-1974: Motivation and Accomplishment*, MIT Press, Cambridge, Mass., Ch. 2.
(b) Johns, B. L. (1977), Importing technology – the Japanese experience and its lessons for Australia, *Sharpening the Focus* (ed. R. D. Walton), School of Modern Asian Studies, Griffith University, Brisbane 4111, pp. 172–84.
21. (a) Rapp, W. V. (1975), Japan's industrial policy, in: *The Japanese Economy in International Perspective* (1977), (ed. I. Frank), Johns Hopkins University Press, Baltimore, Ch. 2.
(b) Tisdell, C. (1977), Japanese and Australian science policy and technology interchange between Japan and Australia, (ref. 21a) pp. 159–71.
22. Rapp, W. V. (see ref. 21a) pp. 40, 41.
23. Planning Bureau, Science and Technology Agency (1977) *Outline of the White Paper on Science and Technology: Facing New Trials in Technological Development*, Japan (unofficial translation (February 1978) Foreign Press Centre), pp. 17, 18.
24. Planning Bureau (see ref. 23) p. 27.
25. See, for instance, Environment Agency (1976), *Quality of the Environment in Japan 1976*, Japan.
26. Rapp, W. V. (see ref. 21a) p. 41.
27. Boltho, A. (1975), *Japan: An Economic Survey 1953-1973*, Oxford University Press, London, p. 136.
28. The Council for Science and Technology (1977), *Science and Technology Policy in the Age of Natural Resources Scarcity*, Report No. 6, Science and Technology

Agency, Japan (unofficial translation by Foreign Press Centre), p. 18.
29. *Ibid.* (ref. 28) pp. 33–40.
30. *Ibid.* (ref. 28) p. 40.
31. *Ibid.* (ref. 28) p. 40.
32. *Ibid.* (ref. 28) p. 21.
33. *Ibid.* (ref. 28) pp. 77–80.
34. *Ibid.* (ref. 28) pp. 62–65.
35. OECD (see ref. 166) Part 6.
36. OECD (see ref. 35) pp. 314–15.

United Kingdom

37. OECD (see ref. 5) p. 15.
38. Lord Privy Seal (1972), *Framework for Government Research and Development*, Cmnd Paper 5046, July, HMSO, London, p. 4. The Lord Privy Seal has the rank of Minister and is attached to a Cabinet Office.
39. For further details and charts of government organization of British science and technology, see, for instance, The British Council (1976), *Government Organisation of Science and Technology in Britain*, London.
40. House of Commons (1976), *Third Report from Select Committee on Science and Technology*, HMSO, London, pp. 86, 87.
41. Lord Privy Seal (1971), *A Framework for Government Research and Development*, November, HMSO, 1971. Includes a report by Lord Rothschild, Head of the Central Policy Review Staff 'The Organisation and Management of Government R & D'.
42. Lord Privy Seal (see ref. 38) pp. 3, 4.
43. The British Council (see ref. 39) p. 4.
44. House of Commons (see ref. 40) p. 87.
45. *Ibid.* (see ref. 44).
46. See *The OECD Observer*, March, 1979, p. 12.
47. For some details of British policies for the stimulation of innovation in industry, see OECD (ref. 16b) Part 7.
48. National Research Development Corporation, *28th Annual Report and Statement of Accounts 1976–77*, NRDC, London.

United States of America

49. *Science and Technology for Development* (1979), US National Paper for 1979 UN Conference on Science and Technology for Development, Department of State, Washington, DC, p. 8.
50. United States Congress, *National Science and Technology Policy, Organization, and Priorities Act of 1976*, Section 102(b).
51. United States Congress (see ref. 50).
52. National Science Foundation (1978), *Science and Technology: Annual Report to the Congress*, NSF, Washington.
53. NSF (see ref. 52) p. 3.
54. For the criterion of selection see Congress of United States (1979), *OTA*

Priorities 1979: With Brief Descriptions of Priorities and of Assessments in Progress, Office of Technology Assessment, Washington.
55. Subcommittee on Domestic and International Scientific Planning and Analysis of the Committee on Science and Technology (1976), *Review of Federal Research and Development Expenditures and the National Economy: Special Oversight Report No. 7*, US House of Representatives, Washington, p. vii.
56. It should be noted that the Budget submitted annually by the President to Congress, contains within it an R & D budget specifying the funding for R & D within the individual agency budgets. For an outline of such a budget see for example, Greenberg, D. S. (ed.) (1977), *Science and Government Report International Almanac – 1977*, Science and Government Report, Inc. Washington pp. 245–83.
57. The figures for 1979 are budget estimates. See NSF (ref. 52) p. 7.
58. For some details see OECD (ref. 16b) Part 3.
59. See NSF (ref. 52) pp. 46, 47.
60. *Science and Technology for Development* (see ref. 49) p. 10.

Some observations

61. Ronayne, J. (1979), *The Allocation of Resources to Research and Development: A Review of Policies and Procedures* (mimeo), School of History and Philosophy of Science, University of New South Wales, p. 2. This monograph also reviews the science policies of Germany, Japan, UK and USA.
62. *Ibid.* (see ref. 61) p. 125.
63. This is so despite increased co-ordination and explicit statements about priorities at the level of the Research Councils in the UK.

CHAPTER FIVE

Science and Technology Policies of Small OECD Economies

5.1 Background

Having considered the science policies of some of the larger economies, let us now look at the policies of a sample of industrialized countries with smaller economies. On the whole these economies are more dependent on foreign trade and technology than are the larger economies. For example a comparison of imports as a percentage of GDP (compare Tables 5.2 and 4.2) show that on average these are higher for the small economies under consideration. The review will help answer the following questions: Are S & T policies different in the small economies from those in the large? Are the priorities of the small economies different? Is there a tendency to be more explicit about S & T priorities in small economies? Do the small economies tend to adopt different mechanisms for co-ordinating and directing their S & T policy? For example, do they take advantage of their small size to more often co-ordinate or centrally direct their science policy? Do these countries use their smaller size to increase public participation in the setting of goals for S & T policy given their highly dependent status in the world? To what extent do their priorities reflect common world-wide trends?

Before considering the S & T priorities of selected small OECD economies (namely those of Belgium, Canada, the Netherlands, Sweden and Switzerland), it is useful by way of background to summarize some of the general features of these economies.

Canada has the largest economy in the group as measured by the size of its GDP. The other countries have economies as measured by GDP around half of the size of the Canadian but the Netherlands has the biggest followed by Belgium, Sweden and Switzerland in that order. The economies show somewhat greater divergence in terms of population but their ordering by size is not altered. Some measures of the relative sizes of the economies are given in Table 5.1.

Table 5.1 Some indicators of the relative size of the economies of Belgium, Canada, Netherlands, Sweden and Switzerland – gross domestic product, population and area.

Country	GDP (billion US $) 1979	Population* (thousands)	Area* (1 000 km^2)
Belgium	111.5	9 841	30.5
Canada	222.8	23 499	9976.1
Netherlands	151.8	13 937	41.2
Sweden	103.3	8 278	450.0
Switzerland	94.1	6 337	41.3

* In 1978.
Source Based on figures in *The OECD Observer*, March, 1980.

All of these small economies exhibit a high degree of international dependence, with the dependence of Canada being least and that of Belgium and the Netherlands being most marked. In the case of Belgium and Luxembourg (these countries are in economic union) imports amount to about half of their GDP and in the Netherlands the proportion is over 40%. Population in relation to agricultural land is very high in the Netherlands and Belgium and high in Sweden and Switzerland. Some indicators of national dependence are given in Table 5.2.

Canada shows the least industrialization of any of the countries in this

Table 5.2 Some possible indicators of international dependence – population density, population in relation to agricultural area, imports as a percentage of GDP – Belgium, Canada, the Netherlands, Sweden and Switzerland, 1978.

Country	Inhabitants per km*	Inhabitants per km^2 of agricultural land	Imports (goods only) as a % of GDP
Belgium	323	656	48.2*
Canada	2	37	21.2
Netherlands	338	677	40.4
Sweden	18	221	23.5
Switzerland	153	315	28.0

* Belgium and Luxembourg.
Source Based on figures in *The OECD Observer*, March 1980.

group and Switzerland and Belgium the most, with the industrial sectors of Sweden and the Netherlands being considerable. Belgium has the smallest proportion of its population engaged in agriculture compared to the other countries with Switzerland having the highest proportion. Switzerland has a markedly smaller tertiary sector than other countries in this group. The structure of civilian employment in Belgium, Canada, the Netherlands, Sweden and Switzerland is shown in Table 5.3.

Table 5.3 Structure of civilian employment by sectors – Belgium, Canada, the Netherlands, Sweden, Switzerland – 1978.

Country	Agriculture forestry and fishing (%)	Industry (mining and manufacturing) (%)	Other (services) (%)
Belgium	3.2	36.6	62.0
Canada	5.7	28.7	65.6
Netherlands	6.2	32.5	61.3
Sweden	6.1	33.0	60.9
Switzerland	8.4	42.8	48.9

Source Based on figures in *The OECD Observer*, 1980.

Gross expenditure on R & D as a percentage of GDP varies considerably between countries in this group. The Swiss percentage is highest followed fairly closely by the Netherlands and then Sweden. On a percentage basis the efforts of these three countries are comparable to those of large economies considered in Chapter 4. GERD as a percentage of GDP is much less for Belgium and Canada than for the other countries. Nevertheless, Canada's expenditure on R & D is the largest in the group and is closely followed by the Netherlands. Expenditure on industrial R & D is of a similar amount in the Netherlands, Sweden and Switzerland but is considerably less in Belgium and Canada in absolute amount. Some measures of scientific effort in these small OECD economies are given in Table 5.4.

Slightly later figures for GERD as a percentage of GDP are: Belgium 1.3 (1977); Canada 1.0 (1977/78); Netherlands 2.1 (1976); Sweden 1.9 (1977) and Switzerland 3.2 (1977). The pattern is similar to the earlier one. The Belgium and Canadian research intensities remain below those of the Netherlands, Sweden and Switzerland. Intensities in the Netherlands and Sweden are comparable to those in the large economies and the Swiss intensity has risen beyond that in the large economies. Absolute GERD in the Netherlands appears to have moved ahead of that in Canada. The

152 Science and Technology Policy

Table 5.4 Indicators of scientific effort – GERD as a percentage of GDP, absolute GERD, industrial R & D – Belgium, Canada, the Netherlands, Sweden and Switzerland (1975 unless otherwise stated).

Country	GERD as % of GDP	GERD in millions US $	Industrial R & D as defined by OECD in millions US$
Belgium	1.0(76)	(790)*	506
Canada	1.1	1702	681
Netherlands	1.9(76)	1600	938
Sweden	1.8	1367	834
Switzerland	2.2	1197	917

* Rough estimate.
Source *The OECD Observer*, March 1979; OECD (1977/78), *Science Resources Newsletter*, Winter.

difference in research intensities may *partly* reflect the different industrial pattern of the countries. Switzerland, the Netherlands and Sweden are the headquarters of a number of multinational companies in research intensive industries. However, there are other factors to take into account also, so let us now consider science and technology policies in Belgium, Canada, the Netherlands, Sweden and Switzerland in turn. The broad features, just outlined, of these economies will provide a useful backdrop.

BELGIUM

5.2 Articulation and administration of priorities in Belgium

Because of rapid technical and economic change and social development in modern economies, the Belgium government believes that a sound science policy is indispensable. According to Kredietbank:

> To conduct a policy means in the first place to set long-term targets, then to create a general but nevertheless sufficiently specified framework within which coherent decisions in various fields can be taken in order to reach those targets, and finally, to stimulate, co-ordinate and control the execution of the decisions taken.[1]

While science policy officially began in Belgium in 1959 with the creation of the Government Science Policy Commission and the National Science Policy Council, their work was largely confined to policies on university expansion, and it can 'hardly be said that they have conducted a policy as determined above'. It was not until 1968 that a Science Policy Department was established by the Belgian government and stock taken of Belgium's overall scientific effort. The establishment of this Department was partly in response to the advice of the National Science Policy Council and general dissatisfaction with the apparent science lag of Belgium behind comparable European countries such as the Netherlands.

The Science Policy Department in conjunction with various government departments such as Education, Economic Affairs, Agriculture and Public Health draws up an *annual* science budget. Within each department a specific amount is set aside for the purpose of supporting scientific and technological effort, and these allocations reflect government priorities. The government's science budget as drawn up includes much more than government expenditure on R & D. Only about 57% of the total science budget is for R & D, 24% is for higher education and 19% for public services such as for libraries, the Institute of Hygiene and Epidemiology etc.

Table 5.5 Belgian budget programme for science policy, 1965, 1972 and 1977 (percentage allocation by areas).

Category	1965	1972	1977
1. Grants (direct) to universities	46.6	53.7	50.0
2. Indirect finance for basic scientific research	6.2	5.2	5.4
3. Technological research for industry and agriculture	18.9	18.0	18.2
4. Public service science and utility	11.9	10.8	11.2
5. Contributions to international scientific institutions	10.7	5.2	8.9
6. Financial transfers	5.7	7.0	7.3
Total budget*	100.0	100.0	100.0

* May not add to 100% because of rounding.

Source Based on 'Science Policy in Belgium' *Kredietbank Weekly Bulletin*, (1978) (March 17), Table 1, p. 2.

The allocation of the Belgian science budget for 1965, 1972 and 1977 by areas selected by the Department of Science Policy is shown in Table 5.5. The categories in the Table cover the following items:

1. Direct grants to universities to cover education and research.
2. Grants from the National Scientific Research Fund and associated funds mostly to academic researchers.

3. This includes an allowance for government support for 'the Institute for Encouragement of Scientific Research in Industry and Agriculture, the prototype department, advanced industrial technology, nuclear research (Research Centre for Nuclear Energy), the National Institute for Extractive Industries, agricultural research and the various National Scientific Institutions'.
4. This heading allows for the financing of *National Impetus Programmes* (which will be discussed later), the Royal Library, Royal Meteorological Institutes and similar institutions.
5. This allows for Belgium's contribution to international bodies dealing with space research and nuclear research and so on and for its S & T aid to developing countries.
6. Takes account of grants from the Public Health Department for support of university hospitals, subsidies on building loans to universities etc.

The major part of the science budget of the Belgian government is spent in universities and associated institutes (all of category 1, most of 2, part of 6, etc.) and this on the whole has shown an upward trend. By contrast the share for industry and agriculture has remained relatively constant at around 18%. The proportion of the budget for international scientific effort is considerable and has fluctuated. Some commentators believe that international commitments entered into by Belgium in the 1960s unduly reduced the flexibility of Belgian science effort.

Belgium's emphasis on university education is a conscious priority. Kredietbank has said of this:

> By continuing this financing effort the government has taken an important political decision, to the effect that a great number of young Belgians should be given a real opportunity for higher education. This is a favourable situation since the main conditions for technological progress are still a thoroughly educated population and sound basic research.[2]

The wide view of S & T policy taken in Belgium accords with the perspective presented in Chapters 1 and 2 of this monograph. In some countries, Germany for instance, science policy and planning tends to be limited to R & D planning.

5.3 Selected features of Belgian science and technology priorities

Like many other countries, Belgium has become increasingly concerned about the social relevance of scientific effort, and the view has been expressed that R & D effort should be applied in an optimal fashion to help solve the problems of society. In 1976 Prime Minister Tindemans expressed the view that while government policies supporting basic science should leave a great deal of freedom to researchers, other types of research should be subject to greater government guidance so as to reflect the needs of the nation.[3] In Tindemans' opinion, university research in particular should be more closely linked to the priorities of society, and especially the requirements of small and medium-sized firms and industries.[4] Since universities in Belgium receive about half of government funds for research and employ more than half of Belgian research workers, the social relevance of university research is of particular importance.

The Belgian economy is a small open one and the government is responding to changing international trading conditions by varying its science policies. Increased competition from developing countries, the rising price of energy resources and the economic recession have led Belgium to place greater emphasis on R & D of industrial significance. The Belgian government believes that:

> To sustain competition with countries possessing abundant raw materials and energy resources on the one hand, and low-cost countries on the other, the new Belgian industrial policy will have to be increasingly based on a more efficient and systematic utilisation of its science and technology potential.[5]

The development of selected industrial sectors has been accorded priority by the Belgian government as well as some intersector goals such as the efficient use of energy and raw materials, improvement of the environment and the quality of life, better working conditions and the effective organization of social services. Government bodies have been directed to take these priorities into account in providing government support for research projects.

Interest-free loans and/or grants are made available to Belgium industry to develop prototypes of new products, to install new manufacturing processes[6] and to finance industrial and nuclear R & D. Special efforts are being made to increase aid for research and innovation by *small and medium-sized firms*. 'Access to government grants in the area of technological research and prototype development has been made easier for small and medium-sized firms (those with fewer than 500 employees)'.

While a significant share of public expenditure on R & D in the past has been directed to the support of R & D in established Belgian industries such as heavy chemicals, metallurgy, textiles and mining, 'it is expected that, in the future, public grants and subsidies will be directed more selectively towards sectors with a high science and technology content.'[7]

Energy research has had a high priority in Belgium. Belgian expenditure on energy research has exceeded that by Sweden and by the Netherlands for example. A considerable amount of this research was concentrated on nuclear energy. About one-fifth of Belgium public expenditure on R & D during the period 1970–75 was allocated to nuclear energy research and development. As Table 5.6 indicates this proportion has declined slightly since the 1960s.

Table 5.6 Allocation of Belgian government funds by object of research and development (%).

Objective	1966	1970	1975	1976	1977
Nuclear energy	23.4	22.0	18.4	19.3	18.8
Space research	6.4	6.0	4.9	5.6	6.2
Defence	1.4	1.3	0.9	0.7	0.5
Technology (for industry and agriculture)	22.1	22.4	26.6	25.6	23.0
Public health	13.2	14.6	12.7	13.6	12.8
Social and economic infrastructure	1.3	1.5	1.7	1.6	1.7
Society and institutions	13.0	12.4	13.2	13.6	15.5
General knowledge	10.2	11.9	12.7	11.6	12.8
Total*	100.0	100.0	100.0	100.0	100.0

* May not add to 100% because of rounding.

Source Based on 'Science Policy in Belgium', *Kredietbank Weekly Bulletin*, (1978) (March 17), Table 2, p. 3.

Table 5.6 indicates that Belgian priorities appear to have altered only slightly as far as R & D is concerned. Relative allocations of funds are fairly steady. Defence research is very small and declining, but research for society and institutions is up slightly as is general knowledge research. A part of the reason for the rise in this last mentioned component is that it includes energy research other than nuclear and Belgian research in this area has risen. As for R & D for industry, the government allocation for this was up in the mid-1970s but appears to have declined again slightly. In this respect Kredietbank reports:

> In spite of the willingness to spend more on other industrial research [apart from nuclear and civil space], this could not be realized as yet. It has indeed been found that the number of Belgian firms which are able to complete a full process of innovation, i.e. ranging from laboratory research to commercial success is very limited.

For this reason government contracts to support industrial R & D 'never really got off the ground'. The present position appears to be one in which the Belgian government wishes to give greater weight to industry and society's needs in R & D but is constrained by institutional realities in moving very quickly to achieve these priorities.

There has been considerable criticism of the relative degree of support given by the Belgian government to industrial R & D.[9] The proportion is much lower than that given by the governments of most large industrialized countries but in fact this may reflect low defence priorities. Nevertheless, the Belgian government has made more progress in giving support to Belgian industrial R & D than may be apparent from Table 5.6. My calculations[10] indicate that the relative contribution to industrial R & D by the Belgian government has increased considerably since the beginning of the 1970s. If nuclear and civil space research are included, the government funds for industrial R & D relative to funds provided by industry itself were as follows: 1971 (14%), 1973 (13.3%), 1975 (17%), 1976 (21%), 1977 (19.97%). If nuclear research and civil space research are regarded as a part of industrial R & D and included in the calculation, the Belgian government's contribution of funds for industrial R & D relative to that of business firms[11] rose from 32.6% in 1971 to 49% in 1977.

A special feature of recent Belgian science policy is the introduction of *National Impetus Programmes* in research. These special programmes are designed to give a fillip to selected areas of research and programmes must be approved by the Belgian Council of Ministers and all the departments concerned must agree to the programmes. (In the Science Policy Budget Table 5.5, they are included under heading 4). A co-ordinator is appointed for each programme and research may be carried out by a *variety* of institutions. The first programme for instance was on 'Water and the Environment' and other programmes are under way. These include data processing, scientific and technical information documentation, social science studies, an energy programme studying the rational use of energy, conservation of energy, new and improved uses of coal and solar energy and another programme studying the utilization of waste and by-products. Kredietbank has said: 'Although they occupy only a modest place in the total science policy budget programme, namely 1.2 per cent in 1978, the national impetus programmes have produced remarkable results'.[12] As yet the Belgian special research projects are on nothing like the scale for instance in the Federal Republic of Germany. They do, however, provide a comparatively direct way of making science effort respond to perceived societal demands.

CANADA

5.4 **Articulation and administration of priorities in Canada**

Government administration of priorities in Canada exhibits elements of centralization and decentralization. The Ministry of State for Science and Technology (MOSST) plays a limited central co-ordinating and advisory role but has no executive power to change the R & D proposals of individual departments. The role of MOSST is a watch-dog one and one of providing positive advice on general science policy and co-ordination of science effort.

The OECD has said of MOSST:

> Henceforth, it will act as a co-ordinating and advisory agency of modest size which chooses very carefully the problems it works on. The Ministry perceived itself as forming part of the central mechanism for policy formulation, collaborating with the Privy Council, the Treasury Board Secretariat and the large departments having R & D activities, in the preparation of proposals to the Cabinet. The Ministry can review and evaluate the forecast expenditures on R & D and furnish advice to the Treasury Board; this latter however retains sole responsibility for approving the R & D proposals of the various departments.[13]

Each year MOSST publishes detailed information on Federal science policy, expenditures and manpower. The main publications in this regard are *Federal Science Activities* and *Federal Science Expenditures* and *Manpower*. Regular detailed data of this kind is useful and is needed to provide a basis for policy advice.

Each year Government Departments directly and scientific bodies associated with departments through their departments independently make submission to the Treasury Board and the Privy Council on their proposed expenditures including science expenditures for the forthcoming year so that their budgets can be approved or modified by the Government. 'Departments and agencies generally carry out and support scientific activities in direct pursuit of their own objectives.'

The federal government of Canada sees science as a means of achieving national goals rather than an end in itself and it does not therefore have a separate Science Budget. Each department and agency has its own goals and objectives (within the government framework) and S & T expenditure is just one possible policy means for achieving goals and objectives and must compete with alternative means.

However, the Ministry of State for Science and Technology reports:

> Recognizing the need for science and technology to contribute to Canada's well-being, the federal government has given MOSST responsibility for advising the government how science may best be used . . . The advice is transmitted to the government through the minister, who is also a member of Treasury Board. [MOSST plays an important advisory role in the budgetary process.] The MOSST analysts concern themselves with both intradepartmental and interdepartmental issues. They assess the relevance of existing and proposed activities . . . If they find inadequate co-ordination, duplication of effort, or the need for improved management, they may recommend specific improvements or a detailed review of all on-going activities, either within a department or even across several departments involved in a given area of S & T activity. If they identify missing or con-

flicting mandates, they may recommend clearer lines of responsibility; perhaps one department will be assigned the lead role.[14]

MOSST has an influence on the co-ordination and management of S & T resources, but this influence should not be exaggerated. MOSST has an advisory influence only and can be hampered in this role by its lack of manpower and its restricted access to relevant data of departments and agencies.

MOSST has recently said:

> Given that individual departments and agencies normally determine the allocation of their resources, there is clearly a need in cross-departmental application areas for a co-ordinating and managing mechanism to establish priorities among the contributing activities and to provide advice to how resources should be allocated among these activities.[15]

There are at present five interdepartmental science and technology committees for co-ordinating purposes. Three of these are 'voluntary' committees for co-ordination by discussion (have only very indirect effect on the allocation of resources) and two have 'official' status and are 'required to report to Cabinet and the Treasury Board on the utilization of existing resources and to recommend changes of priority for new or existing resources where appropriate'. The official interdepartmental committees can and do influence budgetary decisions. The first of the official interdepartmental committees formed was the Interdepartmental Panel on Energy R & D, whose objective is to develop proposals for an integrated program of energy R & D. An Interdepartmental Committee of Transportation R & D has been recently formed with a similar mandate. The three informal interdepartmental co-ordinating committees cover research in space, oceans and food.

Grants (principally for university research) are under the control of three Councils responsible to three different ministers. These councils are (1) the Natural Sciences and Engineering Research Council associated with the National Research Council (NRC), (2) the Social Sciences and Humanities Research Council and (3) the Medical Research Council. The activities of these bodies have not in the past been co-ordinated. However an inter-Council Coordinating Committee is to be established.

> This committee, chaired by the Secretary of MOSST and reporting to the Minister of State for Science and Technology, will have an advisory and coordinating, but not directive role. It will act as a forum for the consideration of areas of mutual concern and will seek to ensure appropriate coverage of recognized disciplines and interdisciplinary research, as well as to harmonize granting practices.[16]

The co-ordinating role of this committee will be limited.

Canadian federal expenditures on the natural and human sciences by

Table 5.7 Canadian federal expenditure on the natural and human sciences by major funding departments 1978/79.

Department	$ millions
Agriculture	134.7
Communications	52.3
Energy, Mines and Resources	125.5
Atomic Energy of Canada Ltd.	94.4
Environment	309.1
External Affairs	
Canadian International Development Agency	27.6
International Development Research Centre	33.8
Industry, Trade and Commerce	65.7
Statistics Canada	138.6
National Defence	95.9
National Health and Welfare	62.7
Medical Research Council	61.4
Science and Technology	
National Research Council	196.6
Grants and Scholarships Program (NRC)	106.6
Secretary of State	
Canada Council – Social Science and Humanities Research Program	32.2
Transport	44.1
Urban Affairs	8.8
Central Mortgage and Housing	15.6
Total major funders	1605.6
Others	222.4
Total federal science	1828.0

Source Based on *Federal Science Activities 1978–79* (1978), Ministry of State for Science and Technology, Canada, Figure 1.

major funding departments and associated agencies are shown in Table 5.7 but this table only gives a partial picture of the bodies involved in science and technology effort in Canada. Some of the main bodies involved in the organization of support for Canadian S & T are shown in Fig. 5.1.

There are clearly many inputs into the articulation of S & T policy in Canada, a policy which is a by-product of general governmental policy.

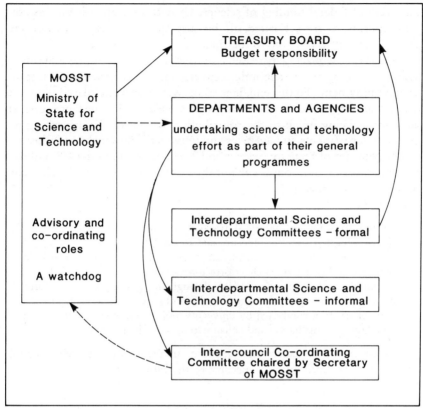

Fig. 5.1 An illustration of some of the federal organizational elements in Canadian support for S & T.

Many departments and agencies and through them interested individuals and organizations have an opportunity to provide an input into the formulation of policy. While there is some co-ordination of S & T policy in Canada and specification of priorities, there is also flexibility and decentralization.

5.5 Selected features of Canadian science and technology priorities

Variations in Canadian S & T priorities reflect worldwide experiences. Canadians during the 1970s placed increased emphasis upon the environment and the quality of life and this was reflected in greater S & T effort in this area. As can be seen from Table 5.7 expenditure on science by the Canadian Department of the Environment in 1978/79 was by far the largest expenditure by any department on science and accounted for about one-sixth of federal funding of science. Growth or emphasis on this area now appears to have tapered off because the economic recession and unemployment have come to the fore.

Canada is now placing increased emphasis on R & D effort as a means of assisting its long-term economic recovery and development, and improving employment. To this end, increasing weight is being placed upon the *contribution* which S & T effort can make *to industry*. For instance, greater attention is being given to the contribution which S & T effort by government departments and agencies and universities can make to industry.

The Minister of State for Science and Technology, in mid-1978 outlined to the House of Commons the major thrust of Canadian R & D policy and measures to support it.

He outlined:

> measures to stimulate industrial research in Canada, to create jobs for scientists, engineers and technicians, and to provide additional support for university research. . . . Mr. Buchanan stressed that the government would strengthen industrial research efforts through the tax incentives already announced, through direct assistance, through changes in government procurement policies, by encouraging Canadian industry to take advantage of the results of research conducted by university and government scientists, and through close consultation and collaboration with the provinces.[17]

Objectives and measures announced or mentioned by the Minister for Science and Technology in a speech[18] to the House of Commons in 1978 included:

1. As a matter of priority, Canada will aim for GERD of 1.5% of GDP by 1983. (Its existing ratio is a little over 1%.)
2. Canadian industrial R & D is to be given greater support by government procurement practice.
3. Contracting-out of government research needs is to be expanded and increased government finance is to be made available to industry to help it undertake specific research projects for government.[19]
4. Grants are to be made available through Canada Works *to create jobs* for S & T personnel to undertake research projects *in* universities at the request of Canadian firms. This will 'enable smaller firms who may not have research facilities to take advantage of university research capacity. It will increase the interaction between business and the universities, as well as provide job opportunities for highly qualified manpower.' This measure complements others to make better use of highly qualified manpower and provide it with suitable *employment*. The science and technology employment programme (STEP) for instance pays up to $14 000 to cover the salary costs of unemployed university and technical graduates hired by firms to conduct research and development within industry.

5. Federal laboratories are to be made increasingly open to the private sector. The aim is to encourage *greater technology transfer* between government and business.[20]
6. Canadian Patents and Developments Ltd. is to act as a central clearing house for the *transfer* of all technology developed in-house by the government. 'The government has confirmed that government laboratories have as one of their objectives the transfer to industry of technology arising from the conduct of intramural work'.
7. Industrial Research and Innovation Centres (IRICs) are to be established at universities to aid industry, especially small and medium-sized business and private inventors, in the development of new products and technologies. They are to be established on the basis of proposals by the universities and provinces.

    These centres will provide a focus for technical, market, legal and patent advice on new ideas to university researchers and businessmen in the region. They will provide industrial access to university expertise and facilities. The IRICs will also facilitate the movement of research workers from industry to university, and vice versa. They will assist in combining the appropriate marketing, management and financial skills necessary to effect transfer of technology, and to establish an entrepreneurial activity needed to spin-off new business based on technology developed in, or with the assistance of university laboratories.[21]

8. The federal government is to assist in the development of regional research Centres of Excellence.

    One of the main objectives of these centres will be to achieve better integration of government, university and industrial capability. They will be based on the natural and human resources of each area and should assist in the development of the industrial capacity of the region.

9. Strategic grants have been introduced to increase university research into areas of national concern. The additional funds made available are distributed through the Research Councils. NRC for instance has placed emphasis on energy, oceanography and toxicology in distributing its funds.

Canada also has a wide range of continuing measures to support performance of R & D by industry itself. Although the Industrial Research and Development Incentives Program, which subsidized increases in research by Canadian firms, was discontinued in 1978/79 it was replaced by increased tax concessions for R & D effort by industry. The tax revenue foregone as a result of these concessions will be much in excess of the subsidies previously paid under the Program. Nevertheless, the Department of Industry, Trade and Commerce will continue many other subsidy schemes to support science and technological effort in industry. Grants will be available under the Enterprise Development Program to support product development, pre-production design and engineering, productivity studies, and market feasibility and strategy studies. The Defence Industry Productivity Program will continue to finance selected R & D projects by firms and plant modernization when this is considered necessary for sustaining the technological capability of the Canadian defence industry. The NRC continues to provide funds for the salaries of industrial scientists engaged in long-term applied research under the Industrial Research Assistance Program. About 10% of the scientists engaged in R & D in manufacturing industry are employed under this scheme. There are also a number of other measures which provide continuing support to R & D in Canadian industry.[22]

Canadian policy has swung strongly towards stimulating and guiding Canadian S & T effort towards improving the competitive position of Canadian industry in the hope that this will assist long-term economic development and employment. The Canadian government appears to be of the view that 'a strong R & D effort is an essential component of success in an international trading environment which is becoming increasingly competitive.'[23] In pursuing this task Canada is aiming for co-operation between government, universities and industry. A great deal of co-operation is already evident between these sectors. Whether or not Canada's policies on S & T will make a useful contribution to the achievement of Canadian goals, however, remains to be seen.

The Canadian government sees science and technological effort as a means, among others, to help achieve national goals. 'Science and technology are not perceived as ends in themselves but as a means of solving human problems and achieving national goals'. Indeed it is because of this fundamental viewpoint, that the role of MOSST, established in the wake of the Lamontagne Reports,[24] has been limited to co-ordination and advice.

THE NETHERLANDS

5.6 Articulation and administration of priorities in the Netherlands

According to the OECD,

> Science and technology played a part early in the economic development of the Netherlands, and 'Science Policy' was practised there thirty years before the term was coined. As far as policy to stimulate industrial innovation is concerned, the Netherlands, together with the United Kingdom, was one of the first countries to initiate an explicit policy to enhance technological innovation in industry, i.e. to create specific institutions for that purpose. The most striking example of the emergence of this scientific and technological policy geared to industrial innovation was the setting up in 1930 of the 'Nederlandse Organisatie voor Toegepast-Natuurweten-Schappelijk Onderzock', better known as TNO (Netherlands Organisation for Applied Scientific Research).[25]

Nevertheless, in recent years there have been significant developments in science policy in the Netherlands, especially as far as the assessment of priorities is concerned.

The role of the Minister for Science Policy is being extended. Until September 1978 his role was merely one of co-ordinating science policy but he now has a modest amount of research funds at his disposal and most importantly he 'will become involved at an early stage of the R & D budget preparations at the individual ministries. This has not been the practice to date, and the effect is that the individual R & D budgets do not become available until late in the day when there is no chance of changing them'.[26] Furthermore, he has been given the responsibility of drawing up a long-term investment plan for research equipment and buildings fully financed by the central government. The Minister's powers to determine the allocation of research funds have been extended. For instance, he is now responsible for determining the Central Government grant to TNO.

The Minister of Science Policy also prepares the Science Budget which is presented to Parliament as part of the national budget. But Mr. Peijnenburg, the present Minister for Science Policy, has brought attention to the fact that:

> the Science Budget has changed in recent years from a mere statement of the financial situation to an exposition of policy by the Minister for Science Policy to the two houses of the States General. It is also acquiring the nature of a multi-year plan for science in the Netherlands.

Forward medium- to long-term planning of government R & D capital expenditure (plant and buildings) is to take place.

Economic stagnation and the growing shortage of government funds have prompted the Dutch to look more closely at the efficiency of public expenditures. It was pointed out in the *1978 Netherlands Science Budget* that it was consequently decided that:

> the resources available for research and development (as in other areas of government activity) had to be put to the best possible use. The maintenance of the research capacity that had been built up and the maintenance, development and efficient running of the system of structures and consultation procedures created for the purposes of co-ordination became major responsibilities. Now there is also a growing realisation . . . that science policy has an important innovatory and stimulatory role. Not only research workers but all those involved – social groups in the widest sense – should guide research, since Government and society are gradually accepting responsibility along with the research workers themselves. The necessity of setting priorities is also felt very strongly at this time; these and the appropriate methods and criteria will have to be decided upon in consultation with research workers and all those involved.[27]

In the 1979 Science Budget, the main background concern continues to be the economic situation and changes in international competitiveness, and these have caused Dutch policy to focus on the importance of Dutch scientific effort to maintain the international economic position of the Netherlands. The Dutch position, as put by the Minister of Science Policy, is:

> Each country prospers best by exploiting its own specific advantages, and the possession of a large pool of technical and scientific knowledge ranks as just such an advantage. In all highly developed industrialized countries governments are increasingly coming to regard it as their responsibility to guide and control this potential in order to derive maximum benefit to the economy.[28]

Later in his science policy statement the Minister says:

> The highly developed countries are also facing great economic problems and this means that the changes in the international division of labour are taking place against a back-drop of conflicting and opposing interests. One of the instruments available for tackling the problems is science policy. However, there is an added complication that the relationship between a country's economic development and its R & D capacity is not clear. It does indubitably exist but it only becomes perceptible in the longer term. The point is that even at this stage there ought to be a lively interaction between science policy and economic policy, for science policy to be effective in the future. This is the main objective of Dutch policy.[29]

Science priorities in the Netherlands are to be set by representatives of society, the government and researchers themselves. The Dutch government is determined that scientific effort should respond to society's needs and has altered its administration of science policy to ensure this. As will be discussed later, *sector councils* are to play a large role in this new direction of scientific effort.

In the Netherlands in the area most closely related to the government, the government research establishments and the universities are the major performers of research. The largest government sponsored research organization is TNO. It has a large number of research establishments or institutes associated with it. These establishments receive funds from several ministries, undertake contract work for industry, disseminate technological information, foster transfer of technology between government departments and to industry and make their R & D equipment and skills available to industry at commercial rates. There are also a number of research centres and laboratories outside the TNO network. These include The Netherlands Energy Research Centre and the Netherlands Ship Model Basin. It might be noted that TNO earns a substantial share of its income from grants from industry, from commercial contacts with industry and from patenting and licensing its inventions.

Table 5.8 Homogeneous group R & D expenditures allocated to Dutch Government departments, 1978, 1979, 1983 (in millions of guilders and by percentage)*

Department	1978 (million guilders)	(%)	1979 (million guilders)	(%)	1983 (est.) (million guilders)	(%)
Education and Science	444.2	34.6	455.2	35.1	495.0	36.7
Economic Affairs	297.5	23.2	303.5	23.4	333.6	24.8
Agriculture and Fisheries	270.0	21.0	282.8	21.8	270.7	20.0
Health and Environmental Protection	96.1	7.5	86.0	6.6	87.4	6.5
Transport and Public Works	77.5	6.0	77.8	6.0	72.9	5.4
Housing and Physical Planning	55.9	4.4	47.1	3.6	39.2	3.0
Other	41.8	3.6	44.7	3.4	47.6	3.5
Total†	1283.0	100.0	1297.1	100.0	1346.4	100.0

* Does not cover all government expenditure in civil R & D. See below.
† Totals may not add to 100% because of rounding.
Source Based on Minister of Science Policy, *Science Policy in the Netherlands: 1979 Science Budget Summary*, Annex 3, p. 26.

While direct grants to universities are made through the Ministry of Education and Science, an important source of secondary income for the universities is the Organization for Advancement of Pure Research (ZWO). This body comes under the Minister for Education and Science and mainly provides supplementary financing for fundamental scientific research in Dutch universities.

Expenditure on civil R & D by the Dutch government (96% of its R & D expenditure) is at present approximately divided as follows in millions of guilders: homogeneous R & D group (1200); universities (1300); other (300); making a total of 2800. Most of the money in the homogeneous group goes towards meeting the expenses of the national research establishments but most of this is channelled through various ministries. The breakdown of homogeneous group R & D by government departments is shown in Table 5.8.

Table 5.8 appears to indicate a shift in Dutch government priorities away from quality of life considerations in R & D towards applied science and science with business and economic applications. The *relative* funds for the first two ministries mentioned in Table 5.8 are expected to rise by 1983 and those for Health and Environment, Housing and Physical Planning will fall. The funds in real terms available to these last mentioned ministries will be less in 1983 than in 1978. This shift is possibly a consequence of the Dutch primary aim to use science policy to support its economy in view of international changes in the division of labour.

An innovation in the organization of Dutch science policy is the formation of *Sector Councils for Science Policy*. These councils are to be established for selected sectors of research and will consist of government representatives, other users and researchers. It is expected that these councils will help needs to be expressed and preserve a balance between the autonomy of research workers and the satisfaction of social priorities. The most important responsibilities of the sector councils are:

(1) To advise the minister or ministers concerned about research policy as a whole within the relevant sector; which means taking into account research by industry and similar research conducted in other countries.
(2) To make recommendations in the form of a multi-year outline plan which is drawn up on the basis of all policy recommendations, multi-year plans or wishes concerning research to be financed by the government, as can be assembled from the appropriate sources (appropriate sources include ministries, research institutions and organisations social organisations).
(3) To promote consultation between institutions involved in research on co-ordinating research plans and programmes, as part of its advisory role.

Sector councils are to be formed for the following areas drawing as far as possible on existing bodies:

1. Energy;
2. Industrial technology;
3. Agriculture;
4. Development co-operation (R & D assistance to less developed countries);
5. Health care and health protection;
6. Building and housing;
7. Defence.

Steps are being taken at present to set up these councils and the Council of Energy Research, coming under the Ministry of Economic Affairs, may be the first council to come into existence. It might be noted that the user/performer relationship envisaged by the Dutch seems considerably wider than that envisaged in the UK in the Rothschild Report in that it makes allowance for social groups outside of Government.

5.7 Selected features of Dutch science and technology priorities

The structure of Dutch government research priorities and changes in

Table 5.9 Dutch Government expenditure on civil research 1978, 1979, 1983 by area of activity (in millions of guilders).

Area of R & D activity	1978 (million guilders)	1979 (million guilders)	1983(est.) (million guilders)
1. Energy	126	124	148
2. Space research and technology	101	97	71
3. Natural environment	31	31	32
4. Medical research	158	166	189
5. Human environment	159	151	139
6. Agriculture	221	233	218
7. Trade and industry	130	144	145
8. Computer science		7	9
9. Social research and research in humanities	137	139	144
10. Research in general (mostly university research)	1555	1623	1683
11. Other	72	80	148
Total	2690	2795	2926

Source Based on Minister of Science Policy, *Science Policy in the Netherlands: 1979 Science Budget Summary*, Annex 2, pp. 24, 25.

priorities can be appreciated from Table 5.9. This sets out all government expenditure on civil research (most of the Dutch government effort) by main fields of research. It indicates a falling priority for expenditure on (a) space research and technology, (b) natural environment, (c) human environment and (d) agriculture. Most other areas are expected to show some slight increase in funding in real terms by 1983. The percentage rise in R & D for energy purposes is likely to show the greatest relative rise between 1979 and 1983. However, on the whole, expenditures by the Dutch government are expected to be fairly stationary in real terms for the next few years.

Funding for R & D in the Netherlands economy is sluggish. GERD as a percentage of GDP has been almost constant since 1976 at around 2.05%, with industry expenditure amounting to about 1.10% of GDP, government 0.96% and universities 0.44%. Government funding for R & D must be seen against Dutch government economic policy. In its overall economic policy, 'the Dutch Government is of the opinion that employment must be given a boost, above all in the private industrial sector, and this entails a cut in the growth of Central Government expenditure in the coming years'.[31] However, given the importance of the Dutch science effort for the international economic competitiveness of Dutch industry, there are to be only minor cuts in government funding for R & D, cuts on a much smaller scale than in other areas of government expenditure. At the same time, government strategy is to use its limited R & D funds more effectively, by improving research efficiency and paying more attention to priorities.

Despite increased co-ordination and planning of Dutch Government scientific effort, the effort is on a sectorial basis and thus is responsive to the wishes of different social groups and is pluralistic. However, given the recent changes in administration of science policy, the degree of central planning, control and co-ordination of government scientific effort by the Ministry of Science Policy has increased.[32]

SWEDEN

5.8 Articulation and administration of priorities in Sweden

The Swedish Institute has said: 'Swedish research policy may be characterized as (1) sectorized, (2) decentralized and (3) pluralistic'.[33] *Sectorization* means that R & D is one instrument, among others, contributing to the realization of the goals of the different sectors: industrial development, defence, health, environmental protection, education etc. Resources in each sector allocated to R & D have to be weighed against other uses which could be made of them to achieve goals. '*Decentralization* means that decisions implementing research policy are largely taken below Cabinet level, i.e. by the independent government agencies and by the research-performing organs or bodies'.[34] *Pluralism* in this case implies that many different bodies participate in decisions about the allocation of public research funds to performers, even though performers of R & D once they receive funds may have a considerable degree of independence in carrying out their research.

While the Swedish form of government is a constitutional monarchy, the organization of the administrative side of government is rather different to the British practice and this effects the method of allocation of public R & D funds. The ministries tend to be very small and concentrate on basic, broad and long-term policy issues. Below the ministries are the agencies and the boards responsible for detailed formulation and application of policy within the powers and goals assigned to them. In the case of R & D, the actual performers (some of which may be government organizations) are below this again. Thus there tends to be three tiers of administration in the implementation of public science effort in Sweden. In a recent report of

STU (The National Swedish Board for Technical Development) the administrative system in Sweden was summarized as follows:

> A basic characteristic of the Swedish civil service is that it is organised at two separate levels – ministries (department) and agencies or boards (*ämbetsverk* or *styrelser*). The former are primarily responsible for the formulation of policy, the latter primarily for its execution. The division of power between ministries and agencies sometimes makes administrative procedure a little complicated. On the other hand it frees the ministers and the ministries from a lot of day-to-day matters.
>
> The work within the ministries is of five general types: policy planning; activities directed towards Parliament; activities which follow from decisions taken by Parliament; matters where power of decision rests with Government instead of an agency; and finally administrative appeals.
>
> The agencies and boards are sub-ordinated to the Government and have to follow given directives, but normally the agencies are given a large degree of independence. Within the limits given by their instructions and their appropriation directives, the agencies have to take their own decisions.
>
> The great advantage of the Swedish system is that it allows the ministries to remain fairly small units, mainly devoted to forming policy for the future development of the country.[35]

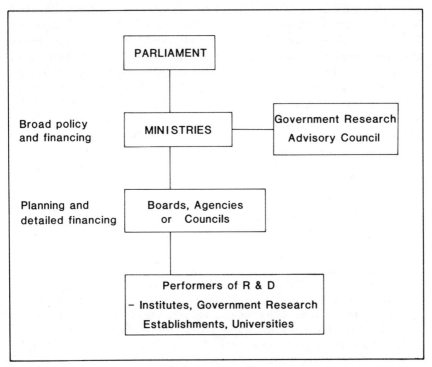

Fig. 5.2 Broad overview of administration of government support for R & D in Sweden.

The broad structure of administration of public support for R & D in Sweden is as indicated in Fig. 5.2. This structure includes the Government Research Advisory Council, under the chairmanship of the Minister of Education. This Council, consisting of Government representatives and members of the research community, advises on the long-term aims of Swedish research policy.

The main ministries allocating public funds to R & D are Education, Industry, Defence, Agriculture and Social Affairs. R & D expenditures by the ministries in 1978/79 and in 1977/78 are shown in Table 5.10. This indicates an increase in relative R & D spending by the Ministry of Industry and by the Ministry of Social affairs and a fall in the relative allocation from the overall budget for Defence R & D – a pattern not unlike the German one discussed earlier.

The Ministry of Education has general responsibility for higher education and the research councils responsible for allocating funds for *basic research*, mostly to universities. Following a report of the Swedish Government Commission on the Organization of Research Councils, these councils were reduced in number and other changes were made. At present there are three research councils: the Humanistic and Social Science Research Council, the Medical Research Council and the Natural Science Research Council. There is also a co-ordinating body, the Council for Planning and Co-ordination of Research (FRN) which works alongside and in co-operation with the councils 'to initiate and finance joint projects in research areas of great social importance'. It is not a superior body and the Commission was careful to recommend that it should not be to reduce possible tensions between the bodies. Since the reforms in 1977 of the research councils membership has been altered to give greater weight to sectorial interests. While, as in the past, the membership of the three research councils consists principally of scientists appointed by academic electoral assemblies, a small number of members now represent sectorial research bodies. 'In contrast, a majority of FRN members are to be representatives of the public interest, such as legislators or MPs.' These changes in composition were designed as a cautious attempt to encourage a better balance between *the societal relevance of research* and its *scientific relevance*. The Commission said:

> The research councils 'cannot ignore the question of societal relevance, and it is a matter of finding forms for dealing with it. This means that the scientific qualifications of a research council organization, and its ability to evaluate research projects from quality aspects must remain unquestioned. In addition it must have a composition and working forms which permit correct assessments of the degree of societal relevance and ensure a balance in the setting of priorities between societal and scientific relevance. The organization must also be able to make the essentially political choice that is necessary as

regards the emphasis in, and the aim of, manifestly societally relevant research'.[36]

The Commission also sets out a number of characteristics which it is desirable for the Council to possess.

There are also special committees associated with the Department of Education concerned with the direction of long-term research and future studies. These are: (1) the Secretariat for Future Studies directed by a board composed of parliamentary members and (2) a Joint Committee for Long-Term Motivated Research associated with the Council for Planning and Co-ordination of Research.

The Ministry of Industry is the Swedish department allocating the second largest amount of its budget to R & D. The largest amount of its funds for R & D are channelled through the National Swedish Board for Technical Development (STU) which in turn allocates these to R & D performers.

Table 5.10 R & D expenditure by the Swedish Ministries 1978/79 and 1977/78 (in millions of Skr).

Ministry	1978/79 (millions Skr*)	(%)	1977/78 (%)
Education	1608	32.0	31.4
Industry	1025	20.4	18.3
Defence	986	19.6	23.4
Agriculture	431	8.6	8.7
Social Affairs	328	6.5	5.8
Transport and Communications	151	3.0	2.8
Foreign Affairs	136	2.7	2.7
Housing	135	2.7	2.1
Budget	118	2.3	2.6
Labour	71	1.4	1.3
Economics	19	0.4	0.4
Local Government	8	0.2	0.2
Justice	7	0.1	0.2
Commerce	4	0.1	0.1
Total	5027	100.0	100.0

* Skr: Swedish kroner.
Source Based on *The National Swedish Board for Technical Development: Tasks–Activities–Structure* (1978), STU, Stockholm, p. 7.

174 *Science and Technology Policy*

Roughly one-quarter of STU's R & D funds go to each of the following: (a) collective research by industrial research associations, (b) R & D in industrial firms, (c) research in universities and institutes of technology and (d) research in government institutions. The Government has assigned three main tasks to STU. These are to:

- promote development of products and systems satisfying needs of society
- increase competitiveness of Swedish industry in world and domestic markets
- increase technical knowledge in general in Sweden.[37]

In order to do this the Government has directed STU:

- to follow technical development by keeping in touch with scientists, institutions and companies,
- to organise and support co-operation in technical research and industrial development and also to encourage contacts between authorities, industry, commerce and research institutions,
- to take initiative on technical research of importance to industry, commerce and society and also to further such research and its utilisation,
- to plan and allocate governmental support through loans and grants to technical research, industrial development work and inventions,
- to follow and monitor the activities of industrial research associations and other co-operative research institutions, at which research work is conducted with governmental support,
- to give advice to inventors and act as an intermediary when it comes to the commercial utilisation of research results,
- to further international technical co-operation with foreign institutions and international organisations.[38]

Associated with STU are the Regional Development Funds, for which substantial sums are provided in the budget. These funds are designed to finance industrial innovation in its later stages (from prototype to test production and market evaluation) and reduce risks to business firms. STU points out that:

> Financing from a Development Fund can help the project through a risky stage. But if the projects turn out to be failures the Funds write off the loans. For successful projects, the loans have to be repaid with interest plus a fee. A principle for a Development Fund is to finance a maximum of 60% of the budgeted development costs, the company or individual carrying the remaining 40%. Being risk-taking bodies, the Funds can also provide capital for investment in production equipment on more normal commercial conditions, and help companies with advice in economic and management questions as well as marketing problems.[39]

As in other countries, Swedish authorities have responded to concern about the national and social environment. Research in environmental conservation and pollution is sponsored by the National Environment Protection Board. Several other public bodies are concerned with R & D into other aspects of the environment. Sweden is also increasing the priority given to assisting less developed countries by means of R & D. The Swedish Institute recently commented:

> As part of the Swedish international development co-operation program, increased assistance is also being given to development research. Part of this assistance is channelled through R & D programs sponsored by international development cooperation agencies and part on a bilateral basis to different development projects. Since 1975 the body responsible for R & D in this field is the Swedish Agency for Research Co-operation with Developing Countries (SAREC), which is administratively linked to the Swedish International Development Authority (SIDA).[40]

5.9 Selected features of Swedish science and technology priorities

Although R & D concerned with social issues has been given greater emphasis, this emphasis is not a first priority. The Swedish Institute states that: 'Even though social issues have been given increased attention in sectorial R & D, efforts in technical R & D still command a high and apparently, according to latest developments, an increasing priority.'[41] R & D of significance to industry has been given increased emphasis.

Sweden has remained active in energy research. In conjunction with the Swedish-based multinational firm ASEA, the Swedish government is attempting to develop nuclear generating plants as independently as possible of foreign technology. In common with most countries, Sweden has recently diversified its energy R & D effort into other energy sources and into energy conservation.

The Swedish government recognizes the importance of international contacts and international R & D co-operation in fostering its objectives. 'Sweden accounts for roughly 1 per cent of the world's total expenditures on R & D. Thus Sweden is extremely dependent on contacts with other countries in research matters.' The Government as a rule includes in the directives to agencies, boards and councils a direction to foster international R & D contact and exchange. While Sweden's first priority is to foster R & D co-operation between Nordic countries, it is also participating in European and wider international R & D arrangements.

While Sweden's expenditures on defence R & D are not on the relative scale of those of the USA and the UK, a substantial share of public funds is used for defence R & D. The share is greater than in most comparable economies. This is a flow-on of Swedish foreign policy. The Swedish Institute has commented:

> Sweden's policy of non-participation in alliances is coupled with an active

defense policy. A massive R & D program is carried on under the aegis of the armed forces and the costs of defense research form a substantial portion of the state's total commitment to R & D. Expenditures continue to be particularly heavy in the field of air force technology. The National Defense Research Institute (FOA), which is the principal defense research organ in Sweden, works in close collaboration with other agencies involved in the total defense. Due to its considerable planning responsibility, FOA undertakes projects of a broader scope than those which ordinarily devolve on a research institute. The costs of the research program for which FOA is responsible amount to about Skr 210 million for fiscal 1978/79.[42]

Despite the absence of a Ministry of Science Policy or equivalent in Sweden and the presence of sectorization, decentralization and pluralism in Sweden, the Swedish science effort appears to be virile and responsive to the needs of society. Possibly the multi-tiered system, involving the agencies, boards and councils at the point below ministerial level, improves the efficiency of the system. The fact too that over 70% of public funds used for R & D are disbursed by only three ministries and that there are a few very large boards (such as STU) or councils involved must assist co-ordination. Within the main sectors mechanisms have grown up to improve co-ordination. New mechanisms are developed when inter-sector co-ordination seems necessary. Take energy for instance:

> Since energy is a subject that affects many societal sectors, special measures have been taken to keep the program in co-ordinated form. The Board for Technical Development, the Council for Building Research, the Delegation for Transportation Research and a newly established Board for Energy Source Development are responsible for different parts of the program which is co-ordinated by the Energy R & D Commission. In addition, special funds have been allotted to the Natural Science Research Council for undertaking efforts in this field. Even though the program comes under the responsibility of several ministries, the responsibility for implementation is under the Minister of Industry. The Energy R & D Commission, which consists of persons representing Parliament, the labor market parties and other organized interests, has the task of collecting a data base to serve as a basis for continued long-term and co-ordinated planning of energy research.[43]

It can be seen that Swedish civil R & D priorities are similar to those at present in the Netherlands but Sweden has not adopted the broad pattern of science policy administration adopted in the Netherlands. Furthermore while it has not adopted the Sector Council of the Netherlands concept, it has increased sector representation on its boards, agencies and councils so that these can take greater account of the societal relevance of the R & D projects which they decide to support.

SWITZERLAND

5.10 Articulation and administration of priorities in Switzerland

While Switzerland by world standards spends a high proportion of its GDP on R & D (over 2.5%), 80% of these funds are provided by industry itself. The fact that only 20% of its R & D funds are provided by the government distinguishes Switzerland from all other developed countries being considered in this review. All the other governments provide a higher proportion of funds for national R & D.

The first priority in Switzerland appears to be the international competitiveness of its industry. 'The consensus is that ability to compete on the world market remains a *sine qua non* necessity for the Swiss economy as well as for the entire Swiss community. Competitiveness is now however generated to a very great extent in research centres and laboratories'.[45] In the case of Switzerland this competitiveness is (as to a lesser extent in Sweden and the Netherlands) generated in the laboratories of a few large multinational companies for which it is fortunate to be the headquarters.

Switzerland is a federation of 25 cantons and constitutional arrangements have affected the development of science policy administration. However, increasing responsibility for higher education and science effort is passing from the cantons to the Confederation. In 1975 for instance, the Federal and Cantonal governments approved measures to co-ordinate their sponsored research programmes and to more efficiently use funds at the higher education level.

The Federal Assembly consists of two Houses:

(a) the Council of States, consisting of members appointed by the cantons;
(b) the National Council, consisting of members directly elected by the citizens.

Every four years the Legislative Assembly elects from its own members a Federal Council consisting of seven members. The Federal Council is the supreme executive body for the Confederation and each of the seven members of the Council, heads a government department.

Both the National Council and the Council of States have Commissions for Science and Research consisting of parliamentary members. The Commission (committee) of the National Council, for example, examines periodically the formulation of overall science and research policy. However a more influential body is *The Swiss Science Council* which advises the Federal Council (the main executive body) 'on all matters of science policy'. It consists of twenty members plus five permanent advisers drawn from government departments. Its membership is such as to involve repre-

178 *Science and Technology Policy*

sentatives from all the main public bodies involved in the support of scientific effort.

Members of the Swiss Science Council are:

> nominated not only for their special qualifications but as representatives of political, business or social groups. This feature distinguishes the Swiss Council from its counterparts in other countries, and underlines its role as a political body whose task is to study problems of science policy or educational policy in such a way that the views of the various parties concerned can be conciliated and action by the Federal Council oriented: the Science Council does not propose programmes drawn up according to strictly scientific criteria but rather submits suggestions based on a political consensus of national community needs and capacities.[46]

The Secretariat of the Swiss Science Council is attached to the Department of the Interior but the Science Council usually reports directly to the Federal Council. It investigates questions put to it by the Federal Council and also undertakes studies on its own initiative. It is an influential body in the setting and filtering of S & T priorities and regularly publishes reports and information bulletins. The Science Council has now come around to the view that the State must be more active in promoting research in Switzerland.

A number of Federal Swiss Government departments are responsible for R & D. The main ones in order of expenditure are Department of Interior, Department of Defence, Department of Public Economy and Department of Transport, Communication and Power. From a science policy point of view, the Department of the Interior is possibly the most important department. It contains within it a Division of Science and Research which provides the administrative body for the Swiss Science Council and the *Swiss University Conference*. This conference body formed in 1968 has a co-ordination role in the federal system of university education.

> The duties of the Conference require it to formulate a largely unified doctrine and promote co-operation. Applications by higher-education establishments for federal grants, which must first be submitted to the Conference, provide such an opportunity. The Conference examines these from the standpoint of intercantonal co-ordination and national educational policy before transmitting them to the Science Council.
>
> The Conference is a co-ordinating organisation rather than an authoritative body. Since it is in this forum that the array of conflicting forces throughout the university sector must be conciliated, a compromise must be found between centralist and cantonal interests.[47]

An institution which has become of increasing importance on the Swiss science policy scene is the *National Fund for Scientific Research*. This Fund was formed in 1952. It is:

a foundation under private law, and thus constitutes an exceptional form of public institution responsible for promoting research. Since all its resources come from the Confederation, the Fund assumes various obligations in regard to the Federal authorities enabling its action to be co-ordinated with the requirements of general science policy.... The foundation must submit yearly for the [Federal] Council's approval a carefully weighed allocation programme by research subjects, and every three years a general report on its plans for encouraging scientific research. Under the Act setting forth these requirements the Federal Council must consult the Swiss Science Council in reaching a decision.[48]

Thus the central advisory role of the Swiss Science Council is preserved.

Three main bodies administer the National Fund: the Foundation Council, the National Research Council and Research Commissions. The Foundation Council consists of a wide range of representatives from different walks of society and 'has been organised so as to allow the voice of all sectors of the Swiss community concerned with the development of scientific research to be heard'. It provides general advice to the National Research Council and appoints its members. The National Research Council is responsible for making research grants and setting out the criteria for grants. The Research Commissions make detailed recommendations to the Research Council on requests received for research grants.

The main function of the National Fund is to promote fundamental research in Switzerland and most of its grants go to university research teams or individuals. It provides a means of supplementary support for basic research in universities. It is also interesting (in view of my comments in Chapter 2) that the National Fund provides publication grants:

> Publication grants enable manuscripts of scientific value to be printed. The National Fund assumes responsibility for printing and negotiating with the publishing firm. The subsidy may take the form of an outright grant in order to reduce the sales price, or a loan which is repayable as sales proceed. The latter has proved effective.... This in fact means that the National Fund guarantees the publisher against slow sales while it receives ample compensation when sales are heavy.[49]

The Swiss National Fund for Scientific Research has, since 1975, become the means of funding a modest target-type national research programme not unlike the programmes undertaken in Belgium and the Netherlands. The Fund:

> has been authorized to devote, during the period 1975–1979, up to 12 per cent of its total budget to financing national research programmes. These are defined as 'concerted actions' aiming at making the best use of existing research potential to obtain results which can be applied in practice to the solution of specific problems, particularly in the socio-economic area. It is target-oriented promotion of scientific research which is expected to build a

bridge between academic research and the research activities of the government bodies or the private sector.

An outline of the programme drafted by the Government is a starting point for a more detailed description of the objectives and stages of implementation, and is intended to serve as a basis for the elaboration of specific research projects. A public tender is then released and the scientific community is invited to bid.[50]

The fact that there is bidding for projects under the Swiss national research programme gives this programme a similarity with the Japanese National Research Programme administered by AIST but the analogy must not be pressed too far. Target-type national programmes give the government an opportunity to express community preferences for research effectively. The first four national programmes selected by the Swiss Government were: (a) prevention of cardio-vascular disease, (b) fundamental water-cycle problems in Switzerland, (c) problems of social integration and (d) energy R & D. Swiss funds available for the national programmes amount to about $6 million annually.

Administrative arrangements now exist in Switzerland which enable broad science priorities to be set. However, there are many customary arrangements and informal contacts which ensure a wide community participation in the formulation of science policy and its goals. In this respect Lucien F. Trueb says:

> A characteristic Swiss feature is that decisions on the allocation of Federal R & D funds are made only after extensive consultations with the Trade and Industry Organization, the so-called Vorort, and major associations of scientists such as the Helvetic Society of Natural Sciences, the Swiss Society for the Humanities, and the Swiss Academy of Medical Sciences. These institutions gather information from a large number of sources in order to define the needs of the country as a whole, to streamline higher education, and to stimulate new and promising avenues of research. However, due to the multifaceted organization of science policies, reaching a consensus is often a long and tedious affair.[51]

The close contact between industry, government and universities in Switzerland (fostered by small cantons or states of the federal system) help to keep all participants aware of the needs of the other. In most universities, there is close contact between local industry and the university. Industry helps fund local university research and members of the business and industrial community not infrequently have part-time university posts. 'The Swiss system' is to some extent different to that of all the other countries studied in this survey.

5.11 Selected features of Swiss science and technology priorities

> Economic competitiveness, as already mentioned on several occasions, is a matter of top priority: the essential aim of research policy must be to establish a scientific and technical environment able to sustain and even stimulate industrial innovation. . . . The principle implies another, arising from the desire to make the most effective use of available resources. Efforts must be so rationalised as to be concentrated in the most promising areas, while any dangerous dispersion of activities must be avoided.[52]

This view is in line with that frequently expressed by the Federal Councillors that Switzerland as a small country must, in order to remain internationally competitive, specialize and find market niches concentrating on a few selected fields in line with the country's scientific and economic potential. The government, however, has not followed a policy of intervention in industrial R & D to promote this goal.

On the whole business has been against government interference in industrial R & D, even though nationally such R & D has a high priority. There is little government subsidy for industrial R & D. The relative lack of demand for, and in some cases opposition to, government intervention reflects:

(a) the desire to remain flexible as far as export markets are concerned; these amount to up to 90% of the market of some industries;
(b) the fact that Switzerland is headquarters for ten very large multi-national companies and that these are in a position to undertake a considerable amount of R & D; these ten companies account for 60% of industrial R & D in Switzerland;
(c) the fact that even small companies are R & D conscious and can co-operate with other sectors such as universities in their research efforts helps; many small firms provide universities and technical institutes with contract research;
(d) 'the country's federal structure, which means that the interest of many groups must be conciliated at national, cantonal and local level and the individualistic character of the Swiss population closely related to the concept of federalised, decentralised powers.'

Even though government direct support for industrial R & D in Switzerland is small, industrial research intensity is high. In 1975 1.7% of industrial resources were spent on industrial R & D, an intensity just slightly less than that of Sweden and the UK and otherwise only surpassed by the USA.[53] However, this intensity has fallen during the 1970s unlike that in Sweden which has been rising markedly. Changes in international competitiveness have had repercussions on Swiss traditional industries such as watch, machine and textile industries. In the last few years temporary financial assistance for R & D in these industries has been granted by the Government.

Like Sweden, Switzerland sets considerable store on international co-operation in R & D effort particularly in big science areas such as nuclear energy. In the absence of international co-operation and sharing of R & D facilities, a small country such as Switzerland would have to abandon some science fields because of its lack of resources.

As pointed out earlier, the allocation of public funding of R & D by socio-economic objective is not a foolproof indicator of national S & T priorities. Nevertheless it provides some indication of priorities. In the case of Switzerland the proportion of total public R & D funding devoted to quality of life objectives (health, environmental protection, social development services, urban and rural planning) was less in 1975 than in all but a few OECD countries. The share for these objectives was much higher in Canada, Sweden, USA, and New Zealand for instance.[54] Furthermore, unlike in most other countries, the share of Swiss public R & D funding for quality of life objectives had already started to fall in the early 1970s. However, in 1976 there was a *major* shift in public funding in favour of quality of life objectives.[55] In particular, *support for health R & D was boosted greatly* and there was more emphasis on social development services. At the same time there was slightly more emphasis on the advancement of knowledge, a major reduction in relative funding for defence R & D and a substantial cut in funding for R & D connected with agriculture, foresty and fishing. Indications are that health R & D is to be made a speciality with *selected* fields being developed with assistance from the National Research Programmes mentioned earlier.

In the early 1970s the OECD said of Swiss Science Policy:

> Scientific activity was long able to develop without seeming to require any strategic framework determined in the light of national needs. Today research is regarded as too costly and too important to warrant anything but choice based on a rational overall strategy.
>
> Such a strategy might be expected to be determined by central government bodies using a largely technocratic approach, and then submitted to the groups for approval. No such procedure is however followed in Switzerland, where the Science Council and the Science and Research Division (of the Department of the Interior) are in the nature of focal points, sparking off the emergence of consensus rather than planning bodies in the controlled-economy sense.
>
> Thus, in association with the representatives of industry, the cantons, the higher education establishments and other Federal departments, these new institutions are motivated by the desire to agree on policies to be implemented. While the quest for agreement may sometimes be long and difficult policies are made easier to carry out owing to the willing co-operation of all parties concerned.[56]

As in Japan (and of course as in some other countries too) national consensus is an important part of decision-making. In common, however, with most other OECD countries faced by stationary public R & D funds, Switzerland has been working towards greater co-ordination and rationalization in the use of available funds. Within the federal framework, this has led to increasing involvement and guidance from the Government of the Confederation.

5.12 Some observations

Changes in science priorities in these small economies have been remarkably similar. During the late 1960s and early 1970s increased weight was placed on improvements in the quality of life (including an improved environment) as an objective but since the energy crisis, world recession and the more obvious impact of changes in the international division of labour, S & T policy for the purpose of increasing the international competitiveness of domestic industry has become of paramount importance. Nevertheless, S & T policy in these countries is not as closely tied in with industrial policy and industrial policy is not as explicitly formulated as in Japan. None of these countries have followed a fully fledged selective industry approach as in Japan although most have moved towards giving preference in funding R & D to selected fields or areas of research chosen in accordance with 'national' priorities.

Governments in these small economies differ in the extent to which they make their overall S & T priorities explicit but all have moved towards increased formulation of aims and making priorities more explicit despite the prevalance of sectorization and decentralization. Possibly the Netherlands has gone further than any other country in this respect. There is now greater feedback between S & T effort in the sectors and collective goals of the societies and more attention is being given to the overall pattern of the S & T policy of the government. Co-ordination of science effort of the sectors is being improved by the availability of greater information, the more frequent use of central clearing bodies for scientific initiatives, greater cross-sector involvement and participation in the formulation of science policies and the use of 'watchdog' or monitoring bodies, such as MOSST in Canada, able to pinpoint major imbalances or anomalies in overall government science effort and bring these to the attention of the public and cabinet for possible action.

The general trend in these small economies appears to be towards greater central co-ordination of government science policy, a trend different from the British one. In no case, however, has a Minister for Science Policy or equivalent been made responsible for directly controlling the whole of the government's science policy effort or a large proportion of government science expenditure. Even in the Netherlands, which appears to have gone further than most countries in its central co-ordination of science policy, the Minister of Science Policy only has modest funds under his direct control and can only exert indirect influence on the science programmes of other government departments in general budget discussions. He does not, for example, have unilateral power to veto the science programmes of individual departments or deny funding for them.

On the question of public participation in the formulation of priorities for science policy and in technology policy, government in all the countries is pluralistic. This means that different pressure groups have a chance to press their point of view through different politicians and government departments and agencies so that a range of views are likely to be represented in the political process of decision-making even at departmental level. The Netherlands, however, has tried to increase public participation in science policy by the appointment of advisory sector councils on which some places are reserved for representatives of the general public. Sweden has also made modest changes in the administration of its science policy designed to give greater weight to public representatives in relation to professional scientist for example on its Council for Planning and Co-ordination of Research (FRN). In all countries there has been increasing pressure on scientists and governments to increase the societal relevance of their research.

Countries such as the Netherlands, Sweden and Canada (not to mention Japan and Germany) appear to take or to be committed to a more systematic approach to science policy than either the USA or the UK have adopted in the recent past. All look upon R & D as an important policy instrument among others to achieve their national goals. On the whole these small economies appear to have remained flexible in their S & T priorities and seem to have reacted promptly to changed world economic conditions. Reaction seems to have been slower in the USA possibly because it is not as dependent on world trade as these smaller economies, recognition lag may be slower, and it may take longer in a larger diverse economy to convince the public and policy-makers of a need for a change in policy direction. As has been emphasized in Switzerland, it is important for S & T policy to remain flexible in small dependent economies since exports account for more than 90% of the market for some industries. The difficult task for any economy is to combine flexibility of policy with resolve and determination of purpose. However, this is necessary if science and technology policy is to be used as an efficient instrument to pursue national goals.

Notes and references

Belgium

1. Kredietbank (1978), Science policy in Belgium, *Weekly Bulletin* (March 17), p. 1.
2. Kredietbank (see ref. 1) pp. 2, 3.
3. See Schuuring, C. (1977), The Netherlands and Belgium, in: *Science and Government Report International Almanac 1977* (ed. D.S. Greenberg), Science and Government Report Inc. Washington, pp. 63–68, particularly p. 67.
4. Schuuring, C. (see ref. 3) p. 67.
5. OECD (1978), *Science and Technology Policy Outlook*, Paris, p. 33.
6. See for instance OECD (1975), *The Aims and Instruments of Industrial Policy: A Comparative Study*, Paris, p. 59.
7. OECD (see ref. 6) p. 47.
8. Kredietbank (see ref. 1) p. 3.
9. See Le rôle de l'industrie en matière de politique scientifique, *Science Technique* (1979), 1st February (4), pp. 437–54.
10. See ref. 9; based on the figures in Table 2, pp. 442
11. For some international comparisons of industrial R & D including Belgium see OECD (1977/78), *Science Resources Newsletter* (Winter).
12. Kredietbank (see ref. 1) p. 4.

Canada

13. OECD (see ref. 5) p. 53.
14. Ministry of State for Science and Technology (1978), *Federal Science Activities 1978–79*, Canada, pp. 62, 63.
15. MOSST (see ref. 14) p. 12.
16. MOSST (see ref. 14) p. 25.
17. Buchanan, J. (Minister of State for Science and Technology) (1978), Support for industrial research announced by the Honourable Judd Buchanan, Press Release, June 1.
18. Buchanan, J. (1978), Measures to strengthen and encourage research and development in Canada, Notes for an Address to the House of Commons, June 1, and 'Highlights' paper tabled.
19. The Department of Supply and Services provides funds on behalf of other government departments and agencies wishing to contract out research.
20. This in part is a flow-on from a report by the Science Council of Canada. See Science Council of Canada (1975), *Technology Transfer: Government Laboratories to Manufacturing Industry*, Report No. 24, Information Canada, Ottawa.
21. Buchanan, J. (see ref. 18) p. 12.
22. For an outline of these see MOSST (ref. 14) pp. 25–28.
23. Buchanan, J. (see ref. 17) p. 3.
24. See Senate Special Committee on Science Policy (Chairman: Honourable Maurice Lamontagne) (1972), *A Science Policy for Canada: Target and Strategies for the Seventies*, Vol. 2, Canada; and (1973), *A Science Policy for Canada: Government Organization for the Seventies*, Vol. 3, Canada.

The Netherlands

25. OECD (1978) *Policies for the Stimulation of Industrial Innovation,* Vol. II–2, Part 8, Paris, p. 132.
26. Minister for Science Policy, (1979), *Science Policy in the Netherlands: 1979 Science Budget Summary,* Information Department, Ministry of Education and Science, The Hague, p. 7.
27. Minister for Science Policy, (1978), *1978 Netherlands Science Budget: Summary,* The Hague, p. 5.
28. See ref. 26, p. 5.
29. See ref. 26, p. 6.
30. Minister for Science Policy (1978), *Sector Councils for Science Policy: Memorandum,* Ministry of Education and Science, The Hague, p. 7.
31. See ref. 26 p. 9.
32. Commencing 1979, the Ministry of Science Policy is to publish a magazine, *Science Policy in the Netherlands* five times a year to provide up-to-date information on Dutch science policy. For a comparatively recent in-depth study of science policy in the Netherlands see *Planning and Development in the Netherlands,* (1976) **7** (2), Van Gorcum Ltd., Assen. This periodical is published by the Netherlands Universities Foundation for International Co-operation and this issue is entirely devoted to Science Policy in the Netherlands.

Sweden

33. The Swedish Institute (1978), Research planning and organization in Sweden, *Fact Sheets on Sweden,* (July) p. 1.
34. See ref. 33, p. 1.
35. *The National Swedish Board for Technical Development: Task-Activities-Structure,* (1978), STU, Stockholm, p. 6.
36. The Swedish Government Commission on the Organization of Research Councils (1975), *Research Councils in Sweden: Proposals for New Organization,* Stockholm, p. 24.
37. STU (see ref. 35) p. 10.
38. STU (see ref. 35) p. 11.
39. STU (see ref. 35) p. 15.
40. The Swedish Institute (see ref. 33) p. 2.
41. The Swedish Institute (see ref. 33) p. 2.
42. The Swedish Institute (see ref. 33) p. 3.
43. The Swedish Institute (see ref. 33) p. 3.
44. Some further references on public administration of R & D in Sweden are: Nordfosk, *Scandinavian Research Guide* and OECD (1978), *Policies for the Stimulation of Industrial Innovation,* Vol. II–2, Part 9, Paris.

Switzerland

45. OECD (1971), *Reviews of National Science Policy: Switzerland,* Paris, p. 17.
46. OECD (see ref. 45) p. 187.

47. OECD (see ref. 45) p. 141.
48. OECD (see ref. 45) p. 163.
49. OECD (see ref. 45) p. 167.
50. OECD (see ref. 5) p. 66.
51. Trueb, L.F. (1977), Switzerland in ref. 3, p. 83.
52. OECD (see ref. 45) p. 201.
53. See *The OECD Observer* (1979), (97, March) p. 12.
54. OECD (see ref. 5) p. 49.
55. See 1976 figure for Switzerland in ref. 5 p. 36. These figures can for instance be compared with the 1975 figures in OECD (1977), *Science Resources Newsletter*, (2, Spring) p. 11.
56. OECD (see ref. 45) pp. 200, 201.

CHAPTER SIX

Retrospect and Prospect

6.1 The increased emphasis on priority assessment in science and technology policy

Because of changing social and economic circumstances, all OECD countries reviewed in this monograph have placed increasing emphasis in their S & T policies upon priority assessment. Community goals are being more carefully formulated especially in so far as they relate to S & T policy and goals are being more carefully ordered and balanced against one another. There is apparently increased determination to get maximum benefit *for society* or the community from public funds spent on R & D and on S & T effort and representatives and members of the community are intervening more than ever to make their wants known and influence the allocation of public R & D funds.

There are a number of reasons for this increased community involvement and concern about S & T policy. Individuals have become increasingly aware of how important S & T change is to their everyday lives and of the extent to which moral and social questions are involved in the application of S & T. The view gained ground in the 1970s that because of these moral and social issues, S & T development and application must be the shared responsibility of all, not merely the 'play toy' or almost exclusive preoccupation of scientists themselves. Concern about environmental issues beginning in the 1960s and gathering momentum by the early 1970s appears to have been the catalyst. Quality of life rather than blind economic growth was set up as a desirable goal and the literature was replete with examples (for instance, use of pesticides, release of heavy metals from factories) indicating the deleterious impacts of S & T. First of all this led to greater social controls on the use of technology and in some quarters had an anti-science backlash, but as thoughts were gathered, communities looked increasingly to science and scientists to help solve environmental problems. Relative expenditures by governments on R & D for quality of life goals rose substantially, and in a few countries are still rising.

However, recently public funding for R & D for quality of life goals has been subject to increased constraints. As a result of the world-wide economic situation, of 'stagflation' (inflation and unemployment), governments in many OECD countries have restrained or reduced public spending with the aim in mind of limiting inflation and stimulating economic activity in the private sector. I do not wish to comment on the effectiveness of this economic strategy here, but wish to observe that public funds are now scarcer in relation to the various goals which the public sector is urged to pursue. Rationally, this means that there are greater pressures for public funds to be used more economically or efficiently to enable wants to be satisfied to the greatest extent possible. Competition for funds between alternative uses in the public sector is now keener and naturally funding for S & T effort has been subject to this increased competitive pressure. In most countries this means that public spending on R & D has been stationary or has declined slightly in the last few years in real terms, although in some countries (Japan principally) efforts are being made to increase public funding of R & D expenditure gradually. Compared to many other activities supported by government, public R & D funding has not been reduced greatly and it seems likely that greater emphasis will be placed on it despite the competition for funds. This is likely to be so because of the energy crisis and the continuing world recession and associated changes in the international division of labour and economic adjustment. Developed countries are increasingly looking to science and technology to assist them in the medium to long term out of their current difficulties.

Although considerable energy research (mostly into nuclear energy) was being undertaken in developed OECD countries prior to the hike in oil prices in 1973 and the onset of the oil crisis, the gravity of the situation was not fully realized for some time. Energy research now has a high priority in all OECD countries and greater emphasis is being placed than previously on other than nuclear sources of energy and on energy conservation and utilization. At the same time, public spending on nuclear energy R & D in developed OECD countries remains high and still accounts for the major part of public funding for energy R & D. The increased social importance of energy research has increased competition for available public R & D funds and has speeded up pressures for setting R & D priorities.

However, the quality-of-life objective has not only had to contend in the last few years with increased pressures from limited public funding and greater priorities for energy research, but with rising demands for public R & D to assist national economic competitiveness and in the long term, employment. Developed OECD countries are experiencing greater competition in traditional manufacturing industries from developing countries and are under pressure to change the structure of their industry and the economic activities in which they specialize. Many developed countries believe that their best strategy is to develop knowledge-intensive industries and to try to maintain their standards of living by keeping ahead of other countries in new technologies and new products which in turn will enable them to boost their export earning. Even here there are priorities to consider.

While quality-of-life objectives still remain a major concern, energy availability and economic competitiveness of national industries are in increased competition as ends for available public R & D funds. With real public funds for R & D stationary or nearly so in most advanced OECD countries and given the apprehended urgency of the above needs by their communities, most countries have accepted the need to allocate their limited public R & D funds in accordance with well formulated priorities.

6.2 Macro approaches to taking account of science and technology priorities

While all countries reviewed in this monograph have a sectoral and a functional approach to science policy (science funding is organized around different sectors and government departments), they differ in the degree of central co-ordination of science activity in the sectors and in the extent to which particular ministries (e.g. in some instance S & T) control a range of research activities which in some countries would be the responsibility of a variety of ministries. In Germany, the Ministry of Science and Technology has extensive co-ordinating power and control over Federal public spending on R & D and in Japan the Science and Technology Agency and the Ministry of Trade and Industry between them have extensive co-ordinating power and control over public funding of national R & D. Close cross co-ordination in science and technology activities and priorities is possible and is being practised in Germany and Japan. In the Netherlands, close cross co-ordination between ministries is now possible given the expanded role of the Minister of Science Policy, and his part in budget deliberations.

Administrative arrangements have been adopted in Canada, Belgium, USA, Switzerland and Sweden to improve cross co-ordination between the S & T activities of different departments and establish broad priorities. Arrangements however remain rather 'loose'. In Canada, for instance, the Minister of State for Science and Technology acts as a science and technology 'watchdog' and plays some role in the allocation of budget funds. In these countries a science budget is drawn up which at least reports the overall expected pattern of public spending on S & T and allows priorities (revealed or otherwise) to be noted and to be debated. In the UK, however, there is little co-ordination of science policy or R & D priorities at the macro-level or intersectoral level. The government has expressed a strong preference for a functional decentralized departmental approach to scientific and technological effort and a science budget reporting the budget of all the departments for science is not published (with the main Budget).

In order to take account of broad and changing community preferences a number of countries (these include Belgium, to a limited extent Canada, Germany, Japan, the Netherlands and Switzerland) have established intersectoral or floating funds for R & D in the form of National Development Programmes. The programmes selected are of a mission type and the selection of projects is usually made by the cabinet or a body representing the interests of a number of sectors. These funds provide flexibility and have been important in promoting co-operation between different types of performers of R & D, e.g. industry, the universities and government.

6.3 Efficiency and science and technology priorities within sectors

At the same time as OECD countries reviewed in this monograph have taken steps to co-ordinate S & T effort between sectors and establish broad overall priorities, most have adopted measures to improve the allocation of public funds for R & D within sectors and increase the efficiency with which funds are used. These measures reflect the growing relative scarcity of available R & D funds in a period when the demand by communities on R & D is rising.

In the UK for example, the recommendations of the Rothschild report have been adopted in an attempt to ensure that research performers meet the needs of customers (government departments) and this requires that departments carefully specify their needs. It is hoped that support for applied research by government departments will be more productive and effective in terms of departmental aims as a result of the customer/contractor approach. In the Netherlands, Sector Councils consisting of scientists and *community* members have or are being appointed for each sector to provide advice to the appropriate minister on R & D needs in the sector as they see them and the efficient use of R & D funds in their sector. Similar organizations exist in Sweden for some sectors and in many countries there is increasing community participation in the setting of priorities.

In some countries it has been possible to make more efficient use of public R & D funds in some sectors as a result of the existence of professional-type management organizations associated with government. In particular, bearing in mind the multi-tiered public administration arrangement, STU in Sweden appears to have been a very effective body in organizing the government's contribution to industrial R & D and applied technical R & D and *following through* with appropriate support in all the stages of

192   *Science and Technology Policy*

applied S & T change. It follows through even to the marketing stage by firms in some instances, even though as a rule it does not bear all the risk but shares it with the innovating company. In the Netherlands, TNO also appears to exhibit a professional management-type approach. Most countries are demanding better assessment of S & T effort. The *US Science and Technology Act* 1976, for instance, recommends that greater use be made of cost-benefit analysis in public decision-making and allocation of funds for the support of S & T effort.

Many of the countries considered in this review have taken steps to improve co-operation between different types of performers of R & D (government, industry and universities) in order to use existing talent and equipment more effectively to meet national priorities and improve technological transfer. This is especially so in Canada, USA, and Japan but is evident in all the other countries reviewed. In particular, attempts are being made to use university R & D resources more effectively to meet perceived community needs.

6.4 Changing science and technology priorities

Within a decade, S & T priorities have changed considerably and this is reflected in the allocation of public funds for R & D. There is now greater emphasis on civil R & D and less emphasis on R & D for defence purposes. Furthermore, public spending on big science has levelled off and even declined in real terms in many OECD countries. Expenditure on basic research and on research in universities has increased slightly in relative terms after having declined slightly.

In the last five years, public funding for energy R & D has increased greatly and diversified in view of the oil crisis and there is greater emphasis on R & D to improve selectively the economic competitiveness of national industries. Emphasis on quality-of-life objectives in S & T effort is now much greater than in the early 1960s, but in many countries competition from the other objectives means that recent emphasis in this area is stationary or declining. In Japan, however, this objective is obtaining greater emphasis. The areas which appear to have been sacrificed in most countries to accommodate these new priorities are defence R & D, space R & D and to a lesser extent agricultural R & D.

Despite the complexity involved in the setting of priorities for S & T, as is for instance apparent from the analysis in Chapters 2 and 3 of this monograph, most industrialized OECD countries have moved towards at least a partial ordering of national S & T priorities and a clearer specification of priorities within sectors in order to use their limited S & T resources more effectively to meet community needs. Concern about the socially optimal allocation of public R & D funds and competition between priorities is likely to intensify in view of the energy crisis, current economic conditions and changing competitive relationships between developed and developing economies. Developed OECD countries are looking to their S & T effort to assist their economies and protect their standing in the world. Even the largest OECD countries realize that they must use their limited S & T resources more wisely if they are to protect their world position. There may also be more fundamental reasons, however, for the trends in government S & T policy.

6.5 Why the trend towards co-ordination and explicit priorities in science and technology policy? Fundamental reasons

A fundamental change has occurred in the science policies of most of the countries studied. There is a strong trend to co-ordination and planning of government science policy and attempts to establish collective and explicit priorities for S & T policy. The rationalist model of policy formulation is being pressed into increasing use in pluralist, functionally orientated and comparatively decentralized government and economic systems. This trend towards centralization is, however, checked by the realization of the bounded rationality of men including those working in government departments and the realization that for the most part S & T advance is not an end in itself but a means, among others, to achieve more basic ends.

In part, the trend to co-ordination in government S & T policies has been made possible by advances in technology itself. Advances in computer technology, modern communication and checking systems and new techniques of assessment have improved the scope for co-ordination and feedback throughout government and indeed the whole economy; but it has also been a response to public demand and other factors.

Public demand for greater control by government over S & T effort stems from many causes. Some fear the risky overspills and threatening possibilities for mankind opened up by science, e.g. nuclear energy. The public generally recognizes that new scientific advances in ill-informed or malicious hands can wreak great havoc on mankind. S & T change has great potential for good as well as evil, the effect of which few or any of us or future generations can be certain to avoid. Demands have been made upon governments to control or limit these threats.

Fundamental moral questions have come to the fore with advances in S & T, questions which earlier generations could more reasonably ignore. Man has so conquered and transformed Nature that it is no exaggeration to say that its future is in his hands. Biological engineering, man's ability to determine the survival of species, the very composition of species on this earth and to alter and endanger his global environment raise great moral questions. How should man use his ability? Would it be better for him not to have such godly powers? The questions raised are so wide-reaching that we cannot ignore them. They effect us all and give rise for a demand for collective or societal priorities in science policy.

The amount of government funding of S & T effort is large and has increased in real terms. On the grounds of accountability for the use of public funds, one would expect increasing pressure to be put upon the government by the electorate to justify such funding and to ensure that government funds are used efficiently to meet community goals. Government bureaucrats can be expected to be favourably disposed to such pressures. Measures to give effect to these demands might be expected to increase the power and influence of public bureaucrats in the community as well as making greater use of their technical skills.

Companies have come to realize more clearly that government S & T policy can in various ways advantage them in international competition and like tariff policy provides one possible way of obtaining a subsidy for their business. Science-based industries in particular can reap substantial rewards or profits from supportive government S & T policies. The benefits of course may flow beyond the companies advantaged by government science policy but it would be naive to consider that such companies lobby for the public interest in proposing science policy changes. Increasingly companies are lobbying for more explicit priorities in government science policy, priorities advantageous to them, ones designed to assist them in their international competitiveness and economic growth. Fortunately, these priorities could also be the ones desired by the wider community but we cannot be certain of that.

6.6 Problems inherent in the basic trend

The desire to establish explicit collective priorities for S & T policy brings some basic difficulties to the fore. How can and should such collective priorities be determined in a community?

If consensus happened to exist about priorities then the problem of establishing collective priorities would appear to be comparatively easy. Nevertheless, they would still need to be distilled from the community and what if the community as a whole is ill-informed? Even the process of distillation could be a difficult one if the revelation of priorities happened to be role-determined. The same person may express a different priority in his role as a company manager, to that in his role as a father and his overall or balanced view (the consensus view) may be different again. If one relied on meetings of company managers and separate meetings of fathers to distill priorities the consensus view might not emerge.

As a rule, however, priorities of different members in the community are not identical. Should collective priorities be established by some amalgam of these individual priorities and what rule should be used to determine the collective ordering? Should we adopt a 'fair' way of doing this and what way is fair? Taking one set of rules for deriving a collective ordering (rules which seem fair) Arrow has shown that it is *impossible* to obtain a social welfare function that satisfies his five rules of fairness![1] No method of amalgamation has yet been devised that will command universal acceptance.

Daniel Bell has said about this matter:

> This problem – of seeking to produce a single social ordering of alternative social choices which would correspond to individual orderings – is academic, in the best sense of the world. In the 'real' world the problem of social priorities, of what social utilities are to be maximized, of what communal enterprises are to be furthered will be settled in the political arena, by 'political criteria' – i.e. the relative weights and pressures of different interest groups, balanced against some vague sense of the national need and the public interest. But it is precisely at this point that the theoretical thorn may begin to prick. For increasingly, one of the issues of a great society – one which can be defined as a society that seeks to become conscious of its goals – is the relationship, if not the clash, between 'rationality' and 'politics'. Much of contemporary social theory has been addressed to the rigorous formulation of rational models of man, in which optimizing, maximizing, and minimizing provide models of behaviour that are rationally normative. But we seem to be unable to formulate a 'group theory' of economic choice. The impasse of social theory, in regard to social welfare, is a disturbing prospect at this stage of the transition to a communal society.[2]

This idealistic quest for the social welfare function appears theoretically (logically) doomed to failure. Furthermore even if the ideal is found, it may be impossible to get individuals to act in accordance with it in the real world. Bell, however, takes some refuge in bargaining as a way to determine collective priorities. He continues:

> I have raised a problem – the lack of an ordering mechanism to make social choices – and quickly taken it to a level of abstraction which is meaningless to practical men. For theorists, the implications are quite drastic, for these logical conundrums strike at the assumptions of those who think that the general will will emerge out of necessity in democratic debate, and those rationalists – as we all may be – who assume that the public interest is discoverable simply by a summation of preferences. Practical men can take heart, for in all this an intuitive idea is reinforced; namely, that differences between persons are best settled, as are so many differences, by bargaining.[3]

The institutions of society, however, help shape bargaining outcomes. This raises the question of how institutions ought to set up to influence bargaining and bargaining outcomes.

Is there a danger that public bodies responsible for formulating science policy will be captured by special interest groups such as business interest groups? Will there be a tendency for the priorities of business to prevail in a bargaining system? A recent OECD report on public decision-making related to S & T observed:

> Government recourse to such advisory mechanisms has traditionally been predicated upon the assumption that the public interest is best served if those whose interests are or might be directly affected by government decisions have an opportunity of shaping those decisions before they are finalised. Such mechanisms have also served as an important means for enhancing governmental technical expertise.
>
> It is, therefore, perhaps not surprising that those groups who feel they have the most to gain (or lose) immediately as a result of certain government decisions or regulations are well represented on such advisory bodies. For instance, nearly 50 per cent of the total membership of the US Federal agency advisory boards is comprised of industrial representatives, with only 7 per cent coming from consumer and environmental groups. One also finds a 'massive over-representation of producers' groups . . . and gross under-representation of all other groups' on the advisory committees attached to Australian Government Departments. While, at the German BMFT, studies have shown a similar representational bias in favour of researchers (55 per cent) and industrialists (25 per cent) on expert commissions responsible for advising the Ministry on science policy matters.
>
> Such limitation in the composition of governmental advisory bodies is often reinforced by the fact that many of them do not provide their members with travel or sitting fees, and thus discourage participation of many people who cannot afford time off from work or whose organisations cannot reimburse them.[4]

Clearly there are dangers that co-ordinating government science bodies can be 'captured' by dominant groups in society and used to promote the ends of these groups more effectively.[5] While greater public participation helps to provide a counterweight to these forces increasingly it takes the form of public protests and demonstration and direct action because of 'disenchantment with the contemporary political processes'.

Is it not also possible that such protests are popular protests against the meritocracy selected to run government and corporations? Bell in forecasting the future of industrial societies suggests:

> The post-industrial society, thus, is also a 'communal' society in which the social unit is the community rather than the individual, and one has to achieve a 'social decision' as against, simply, the sum total of individual decisions which, when aggregated, end up as nightmares, on the model of the individual automobile and collective traffic congestion. But cooperation between men is more difficult than the management of things. Participation becomes a condition of community, but when many different groups want too many different things and are not prepared for bargaining or trade-off, then

increased conflict or deadlocks result. Either there is a politics of consensus or a politics of stymie.

As a game between persons, social life becomes more difficult because political claims and social rights multiply, the rapidity of social change and shifting cultural fashion bewilders the old, and the orientation to the future erodes the traditional guides and moralities of the past. Information becomes a central resource, and within organizations a source of power. Professionalism thus becomes a criterion of position, but it clashes, too, with the populism which is generated by the claims for more rights and greater participation in the society.

If the struggle between capitalist and worker, in the locus of the factory, was the hallmark of industrial society, the clash between the professional and the populace, in the organization and in the community, is the hallmark of conflict in the post-industrial society.[6]

The problems that Bell foresees are of course not peculiar to capitalist societies. The Czechoslovakian writer Radovan Richta, quoted by Bell points out:

There is nothing to be gained by shutting our eyes to the fact that an acute problem of our age will be to close the profound cleavage in industrial civilization which, as Einstein realized with such alarm, places the fate of the defenseless mass in the hands of an educated elite, who wield the power of science and technology. Possibly this will be among the most complex undertakings facing socialism. With science and technology essential to the common good, circumstances place their advance primarily in the hands of the conscious, progressive agents of this movement – the professionals, scientists, technicians and organizers and skilled workers. And even under socialism we may find tendencies to elitism, a monopoly of educational opportunities, exaggerated claims on higher living standards and the like; these groups may forget that the emancipation of the part is always bound up with the emancipation of the whole.[7]

The formulation of collective priorities for S & T effort places many scientists in a quandary. Their scope for independent enquiry is restricted and their autonomy abrogated. When coupled with community funding of science as is increasingly the case in universities for example, creative individualism may be fettered unless it develops in accordance with social priorities. University and other scientists increasingly run the risk of being subjugated to the priorities of others (maybe the populace) and may be increasingly forced into political action to advance their own goals. Collective priorities are a two-edged sword. How can we rationally avoid setting such priorities in an interdependent world of overspills from S & T? Scientists cannot be left free to impinge upon the freedom of others as they please. Is it a delusion, however, to believe that scientists are free and autonomous? Are they merely agents of the dominant managerial/corporate group in society as some neo-Marxists argue?[8] Will the managerial/corporate group dominate the machinery for setting collective S & T priorities in our society if such machinery is set up and compound our present problems? Will such machinery correct the following problem as seen by Gabor:

> If only this army of scientists and technologists could be diverted from technology autonomous, and organized for the good of society! Many thousands of them, of the best and most intelligent, are deeply conscious of the fact that we have got our priorities wrong. The avant-garde of technologists are engaged either in war work, or in pyramid-building (the space race), or they are desperately trying to give something 'to the man who has everything'; to the already overloaded consumer society. They are aware that meanwhile the social machinery is creaking and groaning, that it is racked by pollution, by the senseless drive towards Megalopolis, by stagnation, inflation, unemployment, drug traffic and crime![9]

6.7 Problems inherent in observed government priorities

While expressed and apparent government priorities swung towards quality of life including environmental objectives in the late 1960s and early 1970s and even continued to do so later in some countries, the emphasis now has shifted towards promoting S & T for the sake of increasing the international competitiveness of industry. S & T strategy under government direction is seen as a powerful means to raise exports, raise standards of living, expand employment and meet the challenge emanating from changes in the international division of labour – the gravitation of manufacturing industry to developing countries.

There has been growing acceptance by governments of the product-cycle and monopoly-gains from trade model outlined in Chapter 2 and the desirability of using selective industrial priorities as a means to deal with the immediate problems of unemployment and slower growth than in the past. Rising energy prices have reinforced this trend. How deeply engrained this view has become is clear from the recent recommendations of a Manchester university economist, Stubbs, to Australia, a country grappling with similar economic problems to those of most industrialized countries. Stubbs advises:

> In this study we have sought to show that technological change is a key element in industrial competitiveness and that countries which have not managed to incorporate it fully into their manufacturing industries (e.g. Britain) have performed markedly less well than those which have (e.g. Japan, West Germany). We believe that if Australia is to maintain its employment levels, the existence of an efficient manufacturing sector is essential. . . . There are good theoretical reasons for government to take a leading role in the development and distribution of human capital, and pressing reasons in *realpolitik* why it *must* do so, on a more ambitious scale than formerly. . . . The

Table 6.1 Comparative unemployment rates, rates of inflation and rates of growth of GDP of selected OECD countries (1978, 1979).

Country	Rate of unemployment* (%)	Rate of inflation† (%)	Growth rate‡ (%)
Group A (Relatively low level of unemployment and inflation, higher than average growth rate)			
Japan	2.2	5.8	6
Germany	3.8	5.4	4.25
Sweden	2.2	9.7	4
Netherlands	4.5	4.8	3.25
Group B (Mixed performance)			
Switzerland	0.4	5.1	0.25
Belgium	6.8	5.1	3
Canada	8.3	9.8	2.75
Group C (Relatively high levels of unemployment and inflation lower than average growth rate)			
United States	5.9	13.3	2
United Kingdom	5.5	17.2	2.75

* As a % of labour force 1978.
† Consumer price increases Dec. 78 – Dec. 79.
‡ 1978–1979 %.
Source Based on figures in *OECD Observer*, March 1980.

international transfer mechanisms of technological capacity has grown sharply in its sophistication in the last four decades. Nations that sleep or even doze technologically will waken to find that they have lost opportunities for employment and comparative real incomes lag behind their neighbours.[10]

As we have seen in the reviews of the S & T policies of other OECD countries this theme is more and more apparent, for example in the Netherlands and Sweden, and is likely to become of more significance in the USA where the international industrial competitiveness of many traditional industries such as automobiles has fallen greatly. The current international economic position of the USA exemplifies the Vernon thesis based on the product cycle as reviewed in Chapter 2. Bell has said of this situation:

> American manufactured goods are pricing themselves out of the world market. From the view of theoretical economics, in the inevitable 'product

cycle' of goods production a more advanced industrial society finds itself at a price disadvantage when a product becomes standardized, inputs are predictable, price elasticity of demand is higher, and labor costs make a difference, so that less advanced but competing nations can now make the product more cheaply. And this is now happening in American manufacture. In the world economy the United States is now a 'mature' nation and in a position to be pushed off the top of the hill by more aggressive countries, as happened to England at the end of the first quarter of this century.[11]

It would not be surprising to find increasing American government interest in the realpolitik of using S & T policy to maintain or reduce the slide in the competitiveness of American industry.

Is the recent growing emphasis on using government S & T policy as a means to enhance the international competitiveness of its domestic industry a desirable one? A number of observations appear to be in order. This reaction on priorities is a response to immediate problems of unemployment, inflation and slower growth in industrialized countries. Governments are looking for medium-term solutions to these problems and must appear to the electorate to be attempting to solve these community-wide problems. Increased national industrial competitiveness by giving a technological edge over other nations holds out the national promise of increased employment through greater exports and economic growth, less inflation as a result of greater productivity and higher standards of living through economic growth. It is theoretically possible for all of these ends to be simultaneously achieved. On superficial appearance at least Germany, Japan and Sweden and to a lesser extent the Netherlands have simultaneously been able to achieve these goals, whereas Table 6.1 indicates that the USA and UK in particular have not. Figures for a longer time period only confirm the difference. Largely the difference may be due to the way in which the first mentioned set of countries has been able to integrate S & T with industry and industrial policy, remain flexible and technologically progressive.

Nevertheless, to play at the game of international competitiveness on a grand scale is not riskless. Governments as participants may select the wrong industries and technologies for encouragement or may unwittingly choose the same S & T fields in which to distinguish themselves as do other countries and thus be unwillingly drawn into international cut-throat technological competition. All may lose as a result of this competition and as more countries enter the competition the likelihood of economic loss increases because the development of S & T is not costless. This development requires alternative uses of resources to be foregone.

A basic question is raised: Does the quest for economic growth, full employment and a low rate of inflation through the increased international competitiveness of industry as a result of improved technology offer long-term salvation for mankind from his current and apparently deepening economic problems? Is it possible that increased industrial competitiveness on a *global* scale fed by appropriate S & T policies could increase unemployment and reduce economic growth globally? Labour-saving devices worldwide in a world of inflexible wage rates and relatively inflexible hours of work could increase unemployment and strengthen any existing tendency to underconsumption of production. As a result, companies may intensify their efforts to promote high consumption, for example through advertising. Even if economic growth should be achieved it may be insufficient to restore full employment, may have an adverse impact on the quality of life and environmental conditions and hasten the depletion of non-renewable resources. If a few countries enter the race and others do not take it seriously, the few may gain but if all enter or a large number enter the prospects may be different – the number of losers is bound to rise.

Yet technological progress is not the real villain in this scenario. It is man. It can be argued that man is seeking an easy way out of his current economic problems and is prepared to enter into a Faustian bargain for this purpose. He believes that after all economic growth might still satisfy his dreams at least in the near future even if it brings ecological catastrophe and non-renewable resource depletion closer. In the immediate period, economic growth allows difficult collective decisions about the redistribution of income and the *distribution of work* to be side-stepped, provides hope and reduces immediate resource constraints. Yet if writers such as Daly[12] and Gabor[13] are correct it brings the day of resource crisis closer, the day when difficult decisions can no longer be avoided. Daly argues that the long-term survival of mankind can best be handled by planning for steady-state economies *now*. This means amongst other things setting up institutions to limit the rate of resource depletion and the rate of population growth and to regulate the distribution of income. Whether one agrees or not with the various doomsday philosophies that are now current, it is apparent that promising measures in the shorter period may be disastrous in the long period or worsen economic and social conditions in the long term. Priorities therefore need to be established about the competing claims of the present and the future, including the 'rights' of present and future generations.

In the end the quest for rationality brings us back to a quest for ultimate ends or the ultimate good.[14] Rationality as an ideal must ultimately bring policy-makers and government bureaucrats back to this point despite 'the politics of interest and the politics of passion'. Most of us wish to avoid the possibility of following an insane end rationally or efficiently. There is a danger if we do not think, are content merely to serve or concentrate on narrow priorities, that collectively we could follow insane objectives. This does not mean that ultimate ends are easy to discover or to agree on but it is hopeful that the rational approach brings us face to face ultimately with the need to consider these ends.

Unfortunately this does not mean that if we can obtain agreement about ultimate ends that our fundamental policy difficulties are over. Even though there is agreement, the *individual* self-interest of men and nations may prevent these ultimate ends being realized or acted upon. Individual rationality may conflict with collective rationality and there may be no feasible way to resolve the conflict.[15] For example, let us suppose that we agree with Hermann Daly that a steady-state economy because of the ultimate ends it serves is the most desirable for the *world*. How can such a policy be implemented? Would it be wise for any single nation to adopt the policy alone? In the latter case if an isolated nation adopts a no-economic growth policy and other rapacious countries do not, the no-growth country may be more likely to be captured militarily and economic growth forced on it. Here we have the so-called prisoner's dilemma problem on a world scale.[16] We may all know that collectively we are heading for tragedy but individual self-interest of nations and of individuals may necessitate the tragedy to be played out.

Mishan[17] has suggested that the defence argument for economic growth is not as strong as it is commonly assumed to be. This may well be so but Mishan minimizes the difficulties for maintaining national defence in the absence of economic growth. Mishan suggests that technological progress in defence equipment could be fast or faster in the absence of continuing growth in the output of material goods and it is technology rather than men and materials that are important in modern war. The facility to produce war equipment and develop new technology may show complementarity with a nation's ability to undertake industrial production despite Mishan's suggestion to the contrary.

As far as exports are concerned, the product-cycle thesis suggests that a country may suffer a reduction in export earnings if it is not technologically progressive and growth orientated. This can result in a *fall* in the national standard of living in a country moving towards a steady-state economy. Exports and material standards of living, however, are not ends in themselves. A fall in exports and in material living standards *could* be worthwhile from the point of view of achieving more basic ends.[18]

6.8 In conclusion

In this monograph a swing has been observed in the OECD countries reviewed towards *explicit* priorities in government S & T policy, policy mainly designed for functional purposes. Actual priorities in most countries shifted towards quality of life including environmental objectives in the 1960s and first half of 1970s. Now S & T for increased international industrial competitiveness is being emphasized as an objective, a trend that began in most countries towards the last half of the 1970s.[19] This emphasis appears to be a reaction to alterations in the international division of labour as a result of several developing countries successfully launching production of the traditional manufactures of mature industrial countries, the occurrence of world recession accompanied by unemployment and inflation, reduced economic growth and rising energy costs. Several governments and societies see the strategy of increased international competitiveness of domestic industries encouraged by appropriate government S & T policies as a means to solve unemployment, reduce inflation and increase economic growth. Countries such as Japan and Germany appear to have used such policies successfully. They can work but they are not certain to do so. Furthermore, the more countries that indulge in these policies the greater the chance of these policies not being successful in the world as a whole. They are not explicitly beggar-my-neighbour policies but they could become so in an inflexible economic world. Thus new difficulties for this *realpolitik* strategy could arise on a global scale even ignoring possible adverse long-term effects on the environment, the depletion of resources and the social fabric of society.

It can also be claimed that the widespread demand for explicit national priorities in government policies flows from the growing realization that overspills (in terms of the environment, international industrial competitiveness and otherwise) are of increased significance in interdependent modern industrial economies and require greater attention to be given to collective or societal rationality. The liberal strategy of *laissez-faire*, of sectors of government and corporations blundering along in an uncoordinated or poorly coordinated fashion poses increasing dangers to the community, even though the alternatives to it are not riskless. The quest for greater collective rationality, a trend predicted by Daniel Bell and at odds with the preferred administrative procedures of writers such as Lindblom and Milton Freidman, has required governments to make their S & T priorities more explicit and increase their overall co-ordination of government departments and agencies implementing S & T policy. The drive towards collective rationality has operated both within government sectors and between them, that is at what I have called a micro- and a macro-level. Even in the UK where the general trends to co-ordination between the S & T policies of different government departments and overall priorities has been resisted, the rationality principle has asserted itself at the departmental level since the customer/contractor rule now requires objectives to be made explicit.

The monograph has enabled us to review the specific ways in which a number of OECD countries have responded to underlying societal pressures for greater collective rationality in government S & T decisions and in S & T decisions generally. Technical and theoretical considerations favouring a collective approach to S & T change were reviewed in the first part of the monograph. Some of these considerations such as overspills or externalities from economic activity and the product-cycle of international industrial competitiveness have greatly influenced government S & T policy in the last twenty years as is apparent from the second half of the monograph. The first half of the monograph also enabled questions to be considered about the limits to the efficient administration of public S & T policy and gave us an opportunity to outline trade-offs or choices that cannot be avoided in formulating S & T policy. Technical limits to the use of rational models for formulating and administering government S & T policies were noted.

However, we cannot ignore deeper problems involved in a collective rational approach to policy. How can agreement be obtained about collective ends? Is there any reason to believe that social priorities that do prevail are good or best for society as a whole? Is there a danger of the machinery for setting collective priorities being captured by special interest groups? Problems exist in obtaining agreement about collective ends and ensuring that they prevail. However, even if collective agreement poses no difficulty, can we be sure that the collective ends agreed upon in a society are really good for it? Ultimately, the rational approach drives us to consider what is the good life or what ends are fundamentally worthwhile for society. For many this is a ray of hope. No alternative to the rational approach to policy offers better prospects for humane government because when pushed to its logical conclusion the rational approach brings the uncertainty of ultimate ends to the foreground and thereby emphasizes the need for rational scepticism. Policy and human action must and does go on regardless of whether there is agreement on ultimate ends. Nevertheless, as we proceed we can give consideration to ultimate ends, retain an open mind and attempt to keep at least some of our options open.[20] Nothing is gained in solving our social problems by adopting the stance of an ostrich.

Notes and references

1. Arrow, K. J. (1951), *Social Choice and Individual Values*, John Wiley, New York.
2. Bell, Daniel (1973), *The Coming of Post-Industrial Society: A Venture in Social Forecasting*, Basic Books, New York, p. 307.
3. See ref. 2.
4. OECD (1979), *Technology on Trial: Public Participation in Decision-Making Related to Science and Technology*, Organisation for Economic Cooperation and Development, Paris, p. 56.
5. (a) For a useful review of theories of capture of government regulatory bodies by special interest groups, see Posner, R. A. (1974), Theories of economic regulation, *The Bell Journal of Economics and Management Science*, **5** (2), pp. 335–58.

(b) See also Scherer, F. M. (1980), *Industrial Market Structure and Economic Performance*, 2nd edn, Rand McNally, Chicago, p. 482.
6. Bell, Daniel (see ref. 2) pp. 128, 129.
7. Richta, Radovan (1969), *Civilization at the Crossroads*, (translated from the Czechoslovakian), International Arts and Sciences Press, White Plains, NY, p. 250.
8. (a) The Roses argue, for example, that scientists in capitalist countries for the main part are agents of capitalists. See Rose, Hilary and Steven (eds) (1976), *The Political Economy of Science*, Macmillan, London, Ch. 2.
(b) Mishan also expresses his doubts about the independence of scientists and their idealism. He says 'There is no reason why science and learning should appeal only to those pure in mind and motive' Mishan E. J. (1970), *Technology and Growth: The Price We Pay*, Praeger, New York, Chs. 17 and 18 especially p. 125.
(c) In a more recent book Noble also argues strongly that technology is not an autonomous force in history and that engineers have become agents of corporate capitalism. See Noble, David F. (1977), *America by Design: Science and Technology and the Rise of Corporate Capitalism*, Knopf, New York.
9. Gabor, Dennis (1972), *The Mature Society*, Secker and Warburg, London, p. 44.
10. Stubbs, Peter (1980), *Technology and Australia's Future: Industry and International Competitiveness*, AIDA Research Centre Publication, Melbourne, p. 126.
11. Bell, Daniel (see ref. 2) p. 158.
12. Daly, Herman E. (1979), Entropy, growth and the political economy of scarcity, *Scarcity and Growth Reconsidered* (ed. V. Kerry Smith), Johns Hopkins for Resources for the Future, Baltimore, 1979, pp. 67–94.
13. Gabor, Dennis (see ref. 9).
14. Compare Bell, Daniel (ref. 2) pp. 366, 367.
15. (a) For a discussion of several possibilities for conflict between individual and social rationality see Benn, S. I. and Mortimore, G. W. (eds) (1976), *Rationality and the Social Sciences*, Routledge and Kegan Paul, London.
(b) Also relevant is Rowley, C. K. and Peacock, A. T. (1975), *Welfare Economics: A Liberal Restatement*, Martin Robertson, London, especially Parts 2 and 3.
16. This well known dilemma is outlined for example in Raiffa, H. and Luce, R. D. (1957), *Games and Decisions*, John Wiley, New York.
17. Mishan, E. J. (see 8b) Appendix B.
18. Compare Mishan, E. J. (ref. 8b) Appendix A.
19. For a recent perceptive and critical review of these changing trends see the comments by the poet and conservationist Wright, Judith (1979), in *Science and Technology for What Purpose?* (ed. A. T. Healy), Australian Academy of Science, Canberra, pp. 348–51.
20. It can be argued that rationally this uncertainty means giving more weight to conservation than otherwise.

INDEX

ABRC, UK, see Advisory Board of Research Councils
ACARD, UK, see Advisory Council on Applied Research and Development
Advisory Board of Research Councils (ABRC), UK, 128, 129, 130
Advisory Council on Applied Research and Development (ACARD), UK, 128–31
Advisory science body, 17, 18
Agency for Industrial Science and Technology (AIST), Japan, see AIST
Agency for Small and Medium Enterprises, Japan, 120
Agricultural Research Council (ARC), UK, 127, 129
Agriculture, 90, 97, 151, 153, 154, 168, 169, 182
  and R & D, 192
  and R & D in Canada, 160
Aislabie, C. J., 104
AIST (Agency for Industrial Science and Technology), Japan, 120, 121, 125, 180
Alienation, and scientific and technological progress, 22, 25
Anderson, Kym, 30
Annual Science and Technology Report, USA, 137, 138
Applied research, 130
  balance with basic R & D, 32, 39
  institutes of in Germany, 112
  relative cost in innovation, 92, 93
  in UK compared to Germany and Japan, 132
  in USA, 141, 142
Applied science, 3, 131
  vs basic vs developmental science, 56–59

Appropriation of S & T gains, 4
  neoclassical economic view, 23, 24
Arrow, K. J., 26, 195, 204
ASEA, Sweden, 175
Atomic energy, see also Nuclear energy
  R & D in Canada, 160
Atomic Energy Authority, UK, 129
Atomic Energy of Canada Ltd, 160
Atomic Energy Commission, Japan, 120
Aujac, H., 28
Automation, 97
Australia, 62, 196, 198
  GERD performed by government, 26
  manufacturing industry, 105
  proportion of GERD allocated to higher education, 34
  public provision of scientific services, 66
  technology transfer, 100
    from USA, 103

Baldwin, W., 44, 73
Bargaining, 195, 196
Basic research, 130, 132, 154, 167, 172, 192
  balance with applied R & D, 32, 39, 56–59
  Germany, 114, 115
  in USA, 141–43
Basic science, 131
  vs applied and developmental science, 56–59
  and individual inventors, 101
  and patent system, 5
  related to technology, 78
  UK, 130
Baumol, W. J., 104, 105
Beckermann, W., 14, 21, 27, 28
Belgium, 2, 39, 107, 132, 149, 179, 185, 190, 191, 199

proportion of GERD allocated to higher education, 34
GERD performed by government, 26
GERD size, 151, 152
industrialization of, 151
industrial R & D, 151, 152
international dependence, 149, 150
size of economy compared to Canada, the Netherlands, Sweden and Switzerland, 149, 150
science and technology policies, 152–57
science and technology priorities, 152–57
structure of employment, 151
Bell, D., 197, 199, 203, 204, 205
Benn, S. I., 205
Ben-Porath, Y., 39, 72
Bhagwati, J., 72
'Big push' doctrine of economic development, 21
Big science, 10, 11, 13, 32, 56, 67, 182, 192
and government investment, 7
Japan, 119
UK, 128, 132
Blum, J., 21, 28
BMFT (Ministry of Science and Technology), Germany, 110, 111, 117, 118, 196
role in funding S & T, 112
Boltho, A., 122, 146
Boulding, K., 20
Boxer, A. H., 69, 75
Brazil, new technology and economic development, 70
Bright, J., 28
Britain, *see also* United Kingdom, 198
economic development of, 21
government assistance in diffusion of techniques, 90
patenting in public sector, 92
science for selected industries, 59, 60
transfer of technology from public sector, 92
British Council, 130, 147
British Museum, 127
Buchanan, J., 25, 30, 185
Bureaucracy, and S & T policy, 25
Bureaucrats, 194
Business cycles, due to technological change, 99

Byatt, I., 28

Cabinet, Canada and S & T policy, 158, 159
Cabinet Office, and S & T in UK, 126, 129
Canada, 2, 107, 133, 149, 182, 183–85, 190–92, 199
GERD performed by government, 26
industrialization, 150, 151
industrial R & D, 151, 152
international dependence, 150
preference in R & D grants and science policy for domestic firms, 65, 95
proportion of GERD allocated to higher education, 34
size of economy compared to Belgium, the Netherlands, Sweden, and Switzerland, 149, 150
size of GERD, 151, 152
science and technology policies, 157–64
science and technology priorities, 157–64
structure of employment, 151
Canada Council, 160
Canadian Patents and Developments Ltd, 163
Canadian International Development Agency, 160
Capital, and innovation, 174
requirements and support of German government of R & D, 116, 117
Capital markets, their imperfection and government support for S & T, 7
Capitalism, *see also* Corporate capitalism, 22, 23
Capture theory, 196
Carter, C. F., 34, 38, 39, 71, 72, 73
Carter, J., 138
Centralization, in Canadian science policy, 157
in S & T policy, 15–17, 190, 193
Centres of Excellence, Canada, 163
Centre-periphery theory of economic development, 70
Central Policy Review Staff, and science priorities in UK, 126, 127, 129
Chief Scientific Adviser, UK, 126

Chief Scientists, UK, 126, 130, 131
Childs, G. L., 44, 73
Chile, 95
Civilian R & D, 192
  in USA, 141
Coal utilization research, German, 116
Cohen, A., 27, 28
Collective ends, 204
Collective priorities, 106, 107, 194, 197, 198
Collective rationality, 203
Committee of Chief Scientists and Permanent Secretaries, UK, 126, 131
Community goals and S & T policy, 188
Community participation, *see* Public participation
Community preferences, 180, 191
Communication, and technology transfer, 94
Commission of the European Communities, 146
Committee on Science and Technology (US House of Representatives), 140, 148
Community services and S & T, 10, 11
Comparative advantage, and scientific effort, 42
Competitiveness of industry, *see also* International industrial competitiveness, 144, 194, 198
  and German S & T policy, 114, 115
  in Japan, 125
Computer technology, 193
Congress of United States and S & T policy, 138–40, 147
Consensus, in national goals in Japan, 122
  in S & T policy, 180, 194
Conservation, 205
Consumer society, 198
Contracting-out of R & D, 63, 64
  in Canada, 162
  in Japan, 125
Co-ordination of S & T policy, 15–18, 193
Co-operative research, 64
  in international science, 67, 68
Copyright of publications, and socially insufficient publication, 36, 37
Corporate capitalism, 24
  and scientific and technological change, 22, 29
Council of Energy Research, 168
Council of Ministers, Belgium, and S & T, 157
Council for Ocean Development, Japan, 120
Council for Planning and Co-ordination of Research (FRN) Sweden, 172, 173, 184
Council for Science and Technology (CST), Japan, 120–24, 146, 147
Cox, J. G., 101
CST, *see* Council for Science and Technology
Customer–contractor principle, in British R & D, 130, 131, 144, 191, 203
Cyert, R. M., 27
Czechoslovakia, 197

Daly, H. E., 28, 105, 201, 202, 205
Data processing research, Germany, 113, 115
Davies, M., 103
DAVOR, Germany, 112
Deane, P., 21, 28
Decentralization, and regional distribution of R & D, 62
  in S & T policy, 15–17, 176, 183, 191, 193
  in S & T policy in Sweden, 144
  in S & T policy in USA, 170
Defence, and science policy, 1, 7, 10, 12, 40
  R & D, 32, 53, 54, 192, 202
  R & D in Belgium, 156
  R & D in Canada, 160, 164
  R & D in Germany, 114, 115
  R & D in Sweden, 172, 173, 175, 176
  R & D in Switzerland, 182
  R & D in UK, 126, 127, 129
  R & D in USA, 127, 141, 143
  and transfer of technology, 92
Defence Industry Productivity Program, Canada, 164
Denison, E. F., 21, 28
Department of Agriculture (USDA), USA, 140
Department of Defence (DOD), USA, 139, 140
Department of Health, Education and Welfare (HEW), USA, 139, 140

Department of Education and Science, UK, 126, 127, 129
Department of Industry, UK, and R & D, 129
Department of Science Policy, Belgium, 153
Department of State, USA, 147
Department of Supply and Services, Canada, 185
Developing countries, *see* LDCs
Development, (of an invention), 138
  balance with basic and applied research, 56–59
  government finance for in USA, 141, 142
  government support for, 59
  lags, 79
  relative cost in innovation, 92, 93
Diffusion (of techniques), incentive in France for patented inventions, 89
  and innovation dilemma, 89
Diffusion of new technology, 15, 76, 86–90, 100
  public policy and, 87–90
Diffusion of techniques
  government assistance in Britain, 90
  government assistance in Japan, 90
  and multinational companies, 94
  role of educational institutions, 35, 72
  and technology policy, 31
Diplomacy, and scientific knowledge, 68
Director of OSTP, USA, 136, 137
Division of Science and Research, Department of Interior, Switzerland, 178
Downs, A., 25, 30
Duncan, R. C., 72
Duplication of R & D effort, efficiency and inefficiency, 60, 61
Duplication of S & T effort, a basis for government interference, 7

Eads, G., 39, 72, 102
Ecology, 201
Ecological crisis, and technological change, 105
Economic adjustment, to technological change, *see also* Structural adjustment, 98
Economic bargaining, and scientific knowledge, 68
Economic development, 10, 11, 28
  centre-periphery theory of, 21
  Marxists and neo-Marxists theories, 22
  neoclassical economic theory, 21
  neo-colonial theory, 22
  pessimism about future role of S & T, 22, 23, 29
  role of S & T in, 19–23
  stage and desirable mechanisms of technology transfer, 95
Economic goals for science policy
  in Japan, 122
  in conflict with scientific goals, 58
Economic growth, *see also* Growth of GDP, 13–15, 188, 198–200, 202
  costs of, 20, 21
  as a cure for economic ills, 100
  as a cure for unemployment, 100
  and environmental crisis, 105
  and gains from S & T, 45
  and new technology, 13
  and pollution, 105
  priority in Japan, 118, 119
  and resource crisis, 105
  and stock of knowledge, 38
Economic impact of government S & T policy, 106
Economic recession, 1, 104, 155, 162, 183, 189, 203
  Japanese reaction, 124
Economics of agglomeration of S & T effort, 8
Education, 31, 154
  and inventiveness, 79
  and scientific and technological progress, 21, 32–36
  as a source of technology transfer, 91
  and technological change in society, 98
Educational institutions
  industrial assistance of, 35
  role in diffusion of techniques, 35, 72
EEC (European Economic Community), 10, 12, 13
Efficient set, of imported ideas, 73
Einstein, A., 197
Elitism, 197
Employment, *see also* Unemployment, 55, 169, 198, 200, 201
  argument for basic science, 58

and new technology, 76, 77, 97–99
and replacement of equipment, 90
of university graduates, 36
Ends, 193, 201, 202, 204
Energy, crisis, 183, 189
  efficiency research in Germany, 113, 115, 116
  prospects and science, 32
Energy R & D, 189, 192
  in Belgium, 155
  in Canada, 100, 159, 163
  in Germany, 115, 116
  in Japan, 123
  in the Netherlands, 168, 169
  in Sweden, 175, 176
  reasons for government intervention, 55
  in Switzerland, 180
  in USA, 141, 143
Energy R & D Commission, Sweden, 176
Engels, F., 19, 20, 23, 28
Engineering, 131
Engineers, lack of independence, 205
Enos, J., 79, 101
Enterprise Development Program, Canada, 163
Environment, 1, 21, 32, 119, 157, 188, 193, 198, 201, 203
  and replacement of equipment, 90
  and R & D in Germany, 114, 115
  as an S & T priority in Belgium, 155
  as an S & T priority in Japan, 123
  and technology, 76, 77, 96, 97, 100, 105
Environment Agency, Japan, 121, 146
Environmental controls, restricted by market system, 99
Environmental impact of R & D, 116
Environmental R & D, 167–69, 202
  in Canada, 160, 161
  in Japan, 121
  in Sweden, 174, 175
  in Switzerland, 182
  in USA, 141–43
Equality of income, and technological change, 97
European Organisation of Nuclear Research (CERN), 128
European Recovery Plan (ERP), 117
European Space Agency (ESA), 128
Evans, W. G., 74, 102, 103

Exports, 184, 202
  and new technology, 190
  and science and technology policy, 181, 191
Export of technology, 95, 96
Externalities, see also Overspills, 52, 53, 73, 204
  and technology, 96, 97
External economies, and S & T policies, 8

Federal Co-ordinating Council for Science, Engineering and Technology, USA, 135
Federal Council, Switzerland, 177
Federally Funded Research and Development Centres (FFRDCs), USA, 142
Federal Republic of Germany, see Germany
Feibleman, J. K., 3, 26
Fifth Research Report, Germany, 111, 113
Field, G. M., 103
Firms, see also Small firms
  size of and innovation, 83, 84
  size of and inventiveness, 76, 82, 83
Five-Year Outlook for S & T, USA, 137
Firestone, O. J., 26
Flohl, R., 111, 145
Foreign-owned firms, see also Multinational firms
  as performers of R & D, 32
  and science policy, 65, 66, 71
Foundation Council, Switzerland, 179
France, 56
  economic development of, 21
  proportion of GERD allocated to higher education, 34
  S & T policy, 9, 15
  tax concessions for pollution control, 103
Frank, A. G., 75
Fraunhofer Association, Germany, 112, 113, 117, 118
Freedom, 197
  and science policy, 107
Freeman, C., 42, 72, 83
Functionalism in S & T policy, 129, 191, 190

Gabor, D., 198, 201, 205

Galbraith, J. K., 24, 25, 30, 81, 101
Garvy, G., 104
GDP
  growth as a benefit of technological and scientific progress, 13
  an inadequate measure of welfare, 13, 14
GERD
  Canada, 162
  Germany, 109, 110, 111
  Japan, 109
  Japanese target, 124
  the Netherlands, 169
  optimal level varies with size of country, 39
  optimal percentage of GDP, 38
  percentage financed by governments in OECD countries, 26
  percentage performed by governments in OECD countries, 26
  proportion allocated to higher educational institutions by OECD countries, 34
  in small economies compared to large, 151
  UK, 109
  USA, 109
  Switzerland, 177
German Research Association (DFG), 112, 113
Germany, 107, 133, 144, 145, 154, 157, 172, 184, 190, 191, 196, 198, 199, 200, 203
  economic development of, 21
  GERD allocated to higher education, 34
  GERD compared to Japan, UK and USA, 109, 110
  industrialization compared to Japan, UK and USA, 109
  industrial R & D compared to Japan, UK and USA, 109
  international dependence, 108
  principles of government support of R & D in industry, 116, 117
  R & D effort by socio-economic objective compared to that of Japan and UK, 132, 133
  science and technology policies and priorities, 110–18
  similarity of S & T priorities with those of Japan, 123
  size of economy, 108
  structure of employment, 109
Gibbons, M., 74, 102, 103
Goals for science or technology policy, 9, 13, 25
Government, funding of S & T, 1, 9, 23–25
  vs industry vs universities as performer of R & D, 63, 64
  intervention in S & T effort, 4
  involvement in NRDP in Japan, 125
  S & T policies, 100, 101
Government Science Policy Commission, Belgium, 152
Green, K., 26
Grey Plan, 111
Griliches, Z., 72, 93, 102, 103
Gruber, W., 42, 72, 73
Gusen, P., 73
Gwartney, J. D., 30

Haldane principle, UK, 144
Hamburger, M., 102
Harris, P., 104
Hartley, K., 30, 101
Hauff, V., 146
Haunschild, H., 146
Hay, D. A., 101
Health, and S & T policy, 32, 53, 54
Health R & D, see also Medical R & D, 167, 168, 169, 180
  Belgium, 156
  Canada, 160
  Germany, 114, 115
  Japan, 123
  Switzerland, 183
  USA, 141–43
Helvetic Society of Natural Sciences, 180
Hetman, F., 26, 30
'High civilization', argument for basic science, 58
Higher educational institutions, see also Universities
  compliance with community and economic needs, 35
  and research, 34
Hills, P. V., 103
Hirsch, S., 42, 72
Hirschman, A. O., 27
Hood, N., 103

House of Commons, UK, 147
Hufbauer, G. C., 42, 72
Hughes, Helen, 75
Humanistic and Social Science Research Council, Sweden, 172
Hunter, L. C., 105

Iinuma, J., 72
Imports
  of knowledge, 48–52
  of technology, 93–94
  of technology into Japan, 118
Information technology, 97
Inappropriate technology, in LDCs, 21
  and multinationals, 95
Individual inventors, 76
  and basic science, 101
  decline in importance, 81
  and NRDC, 92
  and public policy, 81
Individualism and science policy, 197
Industry, 191, 192, 194
  benefits from science, 41
  assisted by educational institutions, 35
  vs government vs universities as performer of R & D, 63, 64
  modernization as a priority of German S & T policy, 114, 115
  and NRDP in Japan, 125
  neglect of science for new industries, 61, 62
  and universities, 131, 138, 180
  and university research, 162, 163
Industrial competitiveness, see also International industrial competitiveness, 192
Industrial policy, 183, 200
  and S & T policy, 8, 32, 40–52
  and S & T policy in Japan, 119
Industrial R & D, 151–53, 155, 174, 185
  in Belgium, 156, 157
  in Canada, 162, 163, 164
  principles of German government assistance, 116, 117, 118
  and public policy, 83
  size in Germany, Japan, UK and USA, 109
  in Sweden, 175
  in Switzerland, 181
  in UK compared to USA, Sweden, Switzerland, Germany and Japan, 132

Industrial relations, and new technology, 99
Industrial Research Assistance Program, Canada, 164
Industrial Research and Development Incentive Program, Canada, 163
Industrial Research and Innovation Centres (IRICs), Canada, 163
International scientific effort, Belgium, 154
Industrial technology, Belgium, 154
Inflation, 199, 200, 203
Innovation (in industry), 15, 27, 76, 83–86, 102, 132–34, 137, 142, 164
  in British industry, 147
  defined, 79
  and diffusion of technology dilemma, 73, 86, 89
  diffusion and multinational companies, 94, 95
  influence of market structure and competition, 84–86
  profitability compared to following, 44
  public policy and, 85, 86
  size of firm and, 83, 84
  and technology policy, 31
  willingness to innovate and logistic curve of diffusion of techniques, 88, 89
Institute for Encouragement of Scientific Research in Industry and Agriculture, Belgium, 154
Inter-Council Co-ordinating Committee, Canada, 159
Interdepartmental Committee of Transportation R & D, Canada, 159
Interdepartmental S & T committees, Canada, 159, 161
International affairs, and science, 32, 67–71
  importance of scientific information, 68
International competitiveness, see International industrial competitiveness
International controls on technology, 99
International co-operation, and S & T in Japan, 123
International division of labour, 116, 183, 189, 198, 203

International economic competition, *see* International industrial competitiveness
International industrial competitiveness, 1, 144, 145, 181, 183, 189, 190, 192, 194, 198–204
and Canadian R & D, 164
and German support for R & D, 116
of Japan and S & T policies, 119, 123, 124
of the Netherlands and S & T policies, 166, 170
and STU in Sweden, 174
in Switzerland, 174
International technology transfer, *see also* Technology transfer, 93–96, 100
and multinational companies, 94, 95
International trade, *see also* International division of labour
and gains and losses from new technology, 41
monopoly-gains from, 198
product-cycle and, 198
theories and S & T policies, 42
Invention, 76, 100
cost of marketing, 93
defined, 79
encouragement to be balanced against diffusion, 81
influences on output of, 80, 81
role of individual investor, 81
Inventors, *see* Individual inventors

Japan, 40, 72, 74, 107, 133, 144–46, 183, 184, 189–92, 198, 199, 200, 203
diffusion of techniques to small firms, 90
foreign investment, 95
GERD compared to Germany, UK and USA, 109, 110
GERD compared to USA and USSR, 110
GERD for higher education, 34
GERD performed by government, 26
industrialization compared to Germany, UK and USA, 109
industrial policy, 8, 59, 60, 119
industrial R & D compared to Germany, UK and USA, 109
international dependence compared to Germany, UK and USA, 108
and international trade in technology, 96
main targets of R & D policy, 123
nature of R & D compared to USA, 40
pressure from USA, 119
research institutes to develop industry, 35
socio-economic objectives of R & D compared to those of Germany and UK, 132, 133
science and technology policies and priorities, 118–25
similarity of S & T priorities with German, 123
size of economy compared to Germany, UK and USA, 108
structure of employment compared to Germany, UK and USA, 109
tax concessions for pollution control, 103
and technological and economic forecasting, 18
universities and economic application of research, 36
Japan Research and Development Corporation (JRDC), 120
Japanese National Research Programme, 180
Jevons, F. R., 74, 102, 103
Jewkes, J., 81, 101
Johns, B. L., 146
Johnson, H. G., 58, 74
Johnson, P. S., 64, 74, 84, 101, 102, 103
Joint Committee for Long-Term Research, Sweden, 173

Kahn, A. E., 27, 73
Kaldor, N., 104
Kamien, M. I., 73
Keck, O., 145
Kemp, M. C., 72
Kendrick, J. W., 104
Kennedy, Edwin, 138
King, A., 74, 75
Knowledge, efficient sets of production, 49, 50
import *vs* home production, 48–52
intensive industries, 190
lags in import of, 48, 51, 52
publication, recording, storing and distributing, 36, 179

stock of, 36–38
stock of and economic growth, 38
stock of and S & T progress, 32
Korea, new technology and economic development, 70
Kotler, P., 88, 102
Kredietbank, Belgium, 152–54, 156, 157, 185
Krondratieff cycle, 99, 104
Kuznets, S., 27

Labour movements and technology transfer, 91, 92, 94
Lamontagne, M., 185
Reports, 164
Länder, in German science policy, 111
Langrish, J., 74, 91, 102, 103
Large companies, *see also* Large firms
and gains from export of technology, 96
low research productivity, 81
Large economies, and S & T policy, 106–48
Large firms, *see also* Large companies
and appropriation of gains from international market in new products, 45, 46
and co-operative research, 64
and innovation, 83, 84
inventiveness of, 82, 83
LDCs, 21, 22, 75, 155, 189, 193, 203
Belgian S & T aid to, 154
Dutch assistance, 168
and product cycle, 44
science in development of, 71
and scientific and technical aid, 32, 69–71
Swedish R & D assistance, 175
Learning, by doing, 101
and logistic curve of diffusion of techniques, 88
Leftwich, R. H., 74
Leibenstein, H., 88, 102
Less developed countries, *see* LDCs
Libraries, 33
efficiency in storing knowledge, 37
public provision of, 66
world inter-library loans, 68
Licensing of technology, 95, 96
Lind, R., 26
Lindblom, C. E., 16, 17, 27, 71, 107, 203
Living conditions, as a priority in German R & D policy, 114, 115
Logistic pattern of diffusion of techniques, reasons for, 88, 89
Lord Privy Seal, UK, 129, 130, 147
co-ordinating role in S & T policy, 126
Lowinger, T., 72
Luce, R. D., 205
Luddite fear of technological unemployment, 100
Luxembourg, 150

McConnell, C. R., 73
McDonald, P., 103
Machlup, F., 26
Maddock, Sir Iewan, 131
Malthus, T.R., 19, 20, 28
Management, and innovation, 76, 77, 86
Managerial/corporate group, 197, 198
Managerial inertia, and diffusion of techniques, 87
Managers, and technology transfer, 94, 103
Mandel, E., 23, 29, 30, 75
Manpower planning, in Japan, 125
Mansfield, E., 60, 74, 83, 93, 101, 102
on diffusion of techniques, 88
Market competition, and innovation, 84–86
Market failure
and German government support for industrial R & D, 116, 117
and the environment, 52
and health research, 54
Market mechanisms
compared to political mechanisms, 25
and socially optimal provision of S & T, 24
Market structure, and innovation, 84–86
Marketing
as a cost of innovation, 92, 93
and innovation, 77, 86, 174
networks in technology transfer and diffusion, 93, 94
Market competition, and innovation, 76
Marcuse, H., 22
Marris, R. L., 24, 73
Marshall, A., 27
Marx, Karl, 19, 23, 28
Marxist, *see also* Neo-Marxist, 24
prediction for LDCs, 70
Max-Planck Society (MPG), Germany, 112

Meadows, D. H., 20, 27, 28
Medical research, *see also* Health research, 54, 71, 180
Medical Research Council
  Canada, 159, 160
  Sweden, 172
  UK, 127, 129
Medicine, and technology transfer, 93
Mehta, D., 42, 72, 73
Meritocracy, 196
Minister of Science Policy, the Netherlands, 165, 166, 183, 184, 186, 190
Minister of State for Science and Technology, Canada, 190
Ministry of Education and Science, the Netherlands, 167
Ministry of Science Policy, the Netherlands, 176, 186
Ministry of Finance, Japan, 120, 121
Ministry of International Trade and Industry, *see* MITI
Ministry of State for Science and Technology (MOSST), Canada, 157–61, 162, 163, 183, 185
Ministry of Science and Technology (Germany), *see* BMFT
Mishan, E. J., 20, 27, 28, 105, 202
Mission research, 10
MITI (Ministry of International Trade and Industry), Japan, 119–21, 123, 125, 190
Morality, 193, 197
  and nuclear energy, 104
  and new technology, 97
  and S & T, 188
Monopoly, gains from new technology and trade, 42, 198
  and innovation, 84, 85
Morphett, C., 26
Morris, D. J., 101
Mortimore, G. W., 205
MOSST (Canada), *see* Ministry of State for Science and Technology
Multinational companies or firms, *see also* Foreign-owned firms, 100, 175, 177
  advantages and disadvantages to host country, 95
  as exporters of technology, 95, 96
  and reverse engineering, 103
  and science policy, 65, 66, 71
  in Switzerland, 181
  and technology transfer, 94, 95
Myrdal, G., 28

National Aeronautics and Space Administration (NASA), USA, 139, 140
The National Council, Switzerland, 177
The National Defense Research Institute (FOA), Sweden, 176
National Development Programmes, 191
National Environment Protection Board, Sweden, 175
Natural Environment Research Council (NRC), UK, 127, 129, 131
National Fund for Scientific Research, Switzerland, 178, 179
National Impetus Programmes, Belgium, 154, 157
National Institute for Extractive Industries, Belgium, 154
National Research Council (NRC), Canada, 159, 160, 163, 164
National Research and Development Corporation (NRDC), UK, 92, 93, 129, 133, 134, 147
National Research Programmes, 182
  in Japan, 120, 125
National Science Foundation (NSF), USA, 137, 140, 147, 148
National Science Policy Council, Belgium, 152, 153
National Science and Technology, Organization and Priorities Act of 1976, USA, 135–37, 142, 192
National Scientific Institutions, Belgium, 154
National Scientific Research Fund, Belgium, 153
Natural Science Research Council, Sweden, 172, 176
Natural Sciences and Engineering Council, Canada, 159
National security, *see also* Defence and Defence R & D, 7, 10, 11
The National Swedish Board for Technical Development, *see* STU
Nederlandse Organistie voor Toegepast-Natuurweten-Schappelijk Onderzock, *see* TNO

Nelson, R. R., 26, 39, 72, 102
Neoclassical economics, and
    government support for S & T, 24
    predictions for the future of LDCs, 70
Neo-Marxist, 22, 197
    Frankfurt school, 22
    predictions for the future of LDCs, 70
The Netherlands, 2, 107, 149, 176, 179, 183, 184, 186, 190–92, 198, 200
    GERD, 151, 152
    GERD performed by government, 26
    GERD for higher education, 34
    industrialization of, 151
    industrial R & D in, 151, 152
    international dependence, 150
    size of economy compared to Belgium, Canada, Sweden and Switzerland, 149, 150
    structure of employment, 151
    S & T priorities and policy, 164–70
The Netherlands Energy Research Centre, 166
Netherlands Organisation for Applied Scientific Research, *see* TNO
The Netherlands Ship Model Basin, 166
Neuloh, O., 99, 105
Neutze, G. M., 27
New Left, 24
New products, leadership *vs* imitating, 44
New technology, diffusion of, 86–90
    and foreign trade, 70
    stages in development and use, 80
    theory of international trade, 42, 72
New Zealand, 182
Niskanen, 29, 30
Noble, D. F., 205
Non-renewable resources, *see* Resource
Nordfosk, 186
Nordhaus, W. D., 14, 21, 28, 74
Norris, K., 101, 102
North Atlantic Treaty Organization (NATO), 128
NRDC, *see* National Research and Development Corporation
Nuclear energy, *see also* Atomic energy, 40, 56, 67, 68, 189, 193
    research, 119
    research in Belgium, 154–56
    research in Germany, 113, 115
    research in Japan, 123
    research in Sweden, 175
    and nuclear risks, 1, 6, 7, 97, 104

Oates, W. E., 104, 105
Objectives of research and science, as classified by EEC, 11, 12, 13
    as classified by OECD, 9–11, 13
OECD (Organisation for Economic Co-operation and Development), 9–11, 13, 26–28, 34, 98, 101, 107, 112, 118, 123, 126, 146–49, 151, 152, 158, 164, 185–87, 189, 191–93, 196, 199, 202–3
    larger economies compared, 107–10
    smaller economies, 107, 149–52
    S & T policies and priorities in large economies, 106–48
    S & T policies and priorities in smaller economies, 149–87
Office of Management of the Budget, USA, 136, 144
Office of Science and Technology Policy (OSTP), USA, 135–38, 140, 143, 144
Office of Technology Assessment (OTA), 138, 139
Oil crisis, 192
Oligopoly and innovation, 85
Oppenlander, K., 146
Organization for Advancement of Pure Research (ZWO), the Netherlands, 167
Oshima, K., 27, 40, 72
Over-consumption, 24
Overspills, *see also* Externalities, 116, 203, 204
    and risk in S & T, 26
    and science policy, 52, 53
    from S & T, 24, 193, 197
    and technology, 96, 97
Ozawa, T., 146
Ozga, S., 102

Panel on Energy R & D, Canada, 159
Parker, J. E. S., 77, 101–3
Parliamentary Select Committee on Science and Technology, UK, 126, 128, 129
Patenting of government inventions, 92
Patents, 42, 59, 117
    and co-operative research, 64

Patent system, 5, 26, 36–39
  a dilemma, 89
  in France, 89
Pavitt, K., 7, 8, 26, 27, 74, 102
Peacock, A. T., 205
Pearce, D. W., 73, 103
Peijnenburg, 165
Peterson, W. C., 73
Posner, M. V., 42, 72
Performers of R & D, 63–66, 192
  co-operation in Japan, 124, 125
Philippines, new technology and economic development, 70
Planned Programme Budgeting (PPB), and national S & T goals, 9
Planning Bureau, Science and Technology Agency, Japan, 146
Planning, and S & T, 15, 26, 106, 107, 135, 137, 193
  of science policy in the Netherlands, 170
Pluralism in S & T policy, 134, 176
  in the Netherlands, 170
  in Sweden, 170
  in USA, 134, 135, 144
Political influences, on S & T policies, 25, 58, 60, 61, 122
  by multinational companies, 95
Political *vs* market mechanisms, 25
Pollution, 14, 116
  control in France, 103
  control in Japan, 103
  and economic growth, 105
  and international relations, 100
  and replacement of equipment, 90
  and science policy, 52, 53
  and scientific progress, 21
  and technology, 1, 13
  and technology policy, 96, 97
Population growth, and scientific progress, 20, 23
Post-industrial society, 196, 197
Poverty, and regional distribution of R & D, 62
President's Committee on Science and Technology, USA, 135, 136
Priorities in S & T policies, *see also* relevant OECD countries, 3, 15, 188, 190, 191, 192, 194, 198
  explicit, 193
Priority principle in Japanese S & T policy, 124, 125

Prisoner's dilemma problem, 202
Private inventors, 134
Privy Council, Canada, 158
Product-cycle, 59, 198, 199, 202, 204
  and multinational firms, 95
  and pollution control equipment, 53
  and science policy gains, 71
  theory of international trade, 42–45
Product development, government support of, 39
Product innovation, 43
Prototype department, Belgium, 154
Public participation in S & T policy, 16, 107, 149, 191, 196, 197, 184
  in Japan, 122
Public sector, 134
  and technology transfer, 92–94
Publications, 179
  need for subsidy, 36, 37
Pure science, *see also* Basic science and Basic research, 3

Quality of life, 14, 117, 118–90, 201
  and R & D in Belgium, 155
  and R & D in Canada, 161
  and R & D in Germany, 114
  and R & D in Switzerland, 182
  and S & T policies, 106, 183, 192, 198

Raiffa, H., 205
Rapoport, J., 102
Rapp, W. V., 146
Rationalist model, 193
Rationality, 195, 197, 201–4
  bounded, 193
  and S & T policy, 16, 17, 22, 107
Rational scepticism, 204
Ray, G. F., 102
R & D, 33, 34, 38–40
  in Canada, 158
  contracting-out, 55, 56
  economies of scale, 83
  geographical distribution of, 62, 63
  German public funding of, 110, 111
  optimal percentage of GDP, 38
  performers of, 32, 63
  priorities for government funding in USA, 141, 142
  regional distribution, 62, 63
  relationship to GDP, 38
  in science policy, 31
  spin-off from, 74

in Sweden, 73
R & D effort
  concentration of, 59–63
  distorted under corporate capitalism, 24
  efficiency and inefficiency of duplication, 60, 61, 67, 68
R & D expenditure
  dependence on economic growth rate, 73
  and size of firm, 76, 81, 82
R & D Requirements Boards, UK, 129, 130
Regional Development Funds, Sweden, 174
Regional distribution of R & D, 62, 63
  Japanese policy, 124
Replacement of equipment, 76, 77, 90
  and productivity, 103
Replacement of techniques
  and diffusion of new technology, 86, 87
  reasons for, 79, 80
  Salter-type model, 102
  and science policy, 31
Replacement of technology, 15, 100
Research Advisory Council, Sweden, 172
Research Associations, 64, 174
Research centres of excellence, 63
Research Centre for Nuclear Energy, 154
Research Commissions, Switzerland, 179
Research co-operatives, 117
Research Councils
  Canada, 163
  UK, 148
Research and development, *see* R & D
Research institutes, 117, 132
Research intensity, and size of firm, 82, 83
Rescher, N., 20, 23, 28, 29, 30
Resnick, S. A., 28
Resource depletion, 201, 203
  and economic growth, 105
  and Japanese S & T policy, 122, 123
  and scientific progress, 21
  and S & T policy, 1
Retirement, and technological unemployment, 97
Retraining schemes, 98

Reverse engineering, 44
  and multinational companies, 103
Ricardo, D., 19, 28
Richta, R., 197, 205
Risk, *see also* Nuclear energy, 58, 59, 192, 200
  and German government support for R & D, 116, 117
  and innovation, 133, 134
  in R & D and Japanese policy, 124, 125
  and S & T policy, 5, 6, 55
  and technology, 97, 104
Robertson, D. J., 104
Ronayne, J., 143, 148
Rose, Hilary, 22, 28, 205
Rose, Steven, 22, 28, 205
Rosenberg, J. B., 73
Rosenberg, N., 101
Rostow, W. W., 21, 28
Rothschild customer–contractor principle, *see also* Customer–contractor principle, 130, 131
Rothschild, Lord, 147
Rothschild Report, 130, 168, 191
Rowley, C. K., 205
Royal Library, Belgium, 154
Royal Meteorological Institute, Belgium, 154
Royal Society, UK, 127

Salesmen, importance in technology transfer, 93, 94
Salter, W., 103
Salter-type model, 102
Samuelson, P. A., 20, 28
SAREC (Swedish Agency for Research Co-operation with Developing Countries), 175
Saunders, C. T., 74
Sawers, D., 81, 101
Scherer, F. M., 60, 74, 205
Schmookler, J., 101
Shoeman, M., 28
Schumacher, E. F., 75
Schumpeter, J. A., 29, 81, 84, 101
Schuuring, C., 185
Schwartz, N. L., 73
Science
  distinguished from technology, 3
  for the handicapped, 54, 71
  import of, 50, 51, 71

Index 219

and industrial policy, 40–52
and international affairs, 67–71
neglect of new fields, 61, 62
non-marketability and international co-operation, 67
and social policy, 52–55
link with technology, 76–79
Science budget, 18, 144, 191
  in Belgium, 153, 154, 157
  in Canada, 158
  in the Netherlands, 165
  or R & D budget in USA, 148
Science cities, 63
Science Council of Canada, 185
Science Council, Switzerland, 182
Science policy, *see also* relevant OECD countries
  co-ordination with industrial policy, 32
  and foreign-owned firms, 65, 66, 71
  and health, 53, 54
  options and priorities, 15, 31
  piecemeal, 71
  scope of, 31
Science Policy Department, Belgium, 153
Science Research Council (SRC), UK, 127–28
Science and Research Division, Department of Interior, Switzerland, 182
Science and Technology Agency (STA), Japan, 120, 121, 146, 190
Science and Technology Employment Programme (STEP), Canada, 162
Science and Technology Report, annually in USA, 135
Scientific aid, to LDCs, 69
Scientific effort, concentration of, 59, 71
  and comparative advantage, 42
  returns to, 23
Scientific information and international affairs, 68
Scientific productivity, as a priority in Germany, 114, 115
Scientific progress, 20, 21, 23
Scientific and technological progress
  and education, 21, 33, 34–36
  and stock of knowledge, 33
  and values, 34
Scientists, 188, 191, 197

lack of independence, 205
Secretariat for Future Studies, Sweden, 173
Sector Councils, 166, 168, 176, 184, 191
Sector Councils for Science Policy, the Netherlands, 166, 168
Sectorization, in science policy in the Netherlands, 170
  in S & T policy, 183, 190, 191
  in Swedish science policy, 170
Select Committee on Science and Technology, UK, 126, 128, 129
  concern about customer–contractor principle, 131
Selective policy, industrial, 183, 192
  industrial in Belgium, 155
  industrial in Japan, 119
  industrial in Switzerland, 181
  S & T, 8, 51, 52, 59
  S & T and product-cycle, 183
  S & T in Switzerland, 181
Service science, 66, 67, 71
Shand, R. T., 72
Shrader-Frechette, K. S., 104
SIDA, 175
Silbertson, Z. A., 26, 102, 103
Singapore, new technology and economic development, 70
Skills, and unemployment, 98, 99, 104
Small Business Agency, USA, innovation and, 142
Small economies, 184
  changes in S & T priorities, 183
  disadvantaged in technology export, 96
  GERD of, 110
  and international technology transfer, 100
  S & T policies in, 149–87
Small (and medium) firms
  appropriate little from international trade in new products, 44, 45
  assistance with R & D in Germany, 117, 118
  Belgium, 155
  case for assistance with S & T, 6, 7
  Canada, 163
  and diffusion of techniques, 90
  and innovation, 83, 84
  inventiveness of, 82, 83
  Report of Committee of Enquiry, UK, into, 103

technical assistance in Japan, 120
and transfer of technology, 91
Smith, Adam, 19, 28
Smith, B. R., 27
Snow, C. P., 56
Social impact, of S & T policy, 106
Social goals or priorities for S & T policy, *see also* Societal . . ., 1, 2, 32, 52–55, 106, 107, 197
in Japan, 122
Social Science and Humanities Council, Canada, 159, 160
Social Science Research Council (SSRC), UK, 127, 129
Social services, organization of an S & T priority in Belgium, 155
Social welfare
function, 195
and scientific effort, 40
Sociological aspects of technology, 97–99
Sociological costs of technological change, 99, 100
Socialism, 197
and S & T progress, 22
Societal demands on S & T, 157
Societal goals, and S & T policy, 188
Societal problems, and German science policy, 118
Societal priorities in S & T, 194
Societal rationality, 203
Societal relevance of S & T or R & D effort, 154, 155, 172, 176, 184
Society, structure and technology, 76, 77, 97, 100
Socio-economic objectives
in Japanese S & T policy, 124
in R & D in Germany, 133
in R & D in Japan, 133
in R & D in UK, 133
South Africa, 55
Soviet Union, 20, 22, 110
Space Activities Commission, Japan, 120
Space R & D, 67, 119, 192, 198
Belgium, 154, 156
Canada, 159
Germany, 113, 115
the Netherlands, 169
UK, 128
USA, 141, 143
Spill-overs of technologies, *see also*
Overspills and Externalities, 15
Spin-off, from defence R & D, 55, 56
S & T Five-Year Outlook, USA, to be supplied annually, 135
S & T policies and priorities, *see also* Table of Contents
of large economies, 106–45
of small economies, 149–85
and technological, economic and social forecasting, 18, 19
Staehle, 104
Stagflation, 189
Standard of living, 12, 202
and new technology, 190
and S & T, 12, 13, 198
Steady-state economies, 201
Stieber, J., 104
Stillerman, R., 81, 101
Structural adjustment (of industry), 116, 189
Structural unemployment, and technological change, 100
STU (The National Swedish Board for Technical Development), 171, 173, 174, 176, 186, 191
Stubbs, P., 105, 198, 205
Supersonic transport, 40, 56, 100
Sweden, 2, 39, 99, 107, 132, 149, 177, 181, 182, 184, 190, 191, 199, 200
GERD performed by government, 26
GERD allocated to higher education, 34
GERD size, 151, 152
industrialization of, 151
industrial R & D, 151, 152
international dependence, 150
S & T policies and priorities, 170–76
size of economy of compared to Belgium, Canada, the Netherlands and Switzerland, 149, 150
structure of employment, 151
Swedish Agency for Research Co-operation with Developing Countries (SAREC), 175
Swedish Government Commission on the Organization of Research Councils, 172
The Swedish Institute, 170, 175, 186
Swedish International Development Authority (SIDA), 175
Swiss Academy of Medical Sciences, 180

Swiss National Research Programme, 180
The Swiss Science Council, 177–79
Swiss Society for the Humanities, 180
Swiss University Conference, 178
Switzerland, 2, 39, 107, 132, 149, 184, 186, 187, 190, 198
 GERD size, 151, 152
 industrialization, 151
 industrial R & D, 151, 152
 international dependence, 150
 size of economy compared to Belgium, Canada, Sweden and Switzerland, 149, 150
 S & T policies and priorities, 177–83
 structure of employment, 151

Taiwan, new technology and economic development, 70
Taylor, C. T., 26, 102, 103
Technocrats, 24
Technological change
 and ecological crisis, 105
 as a scapegoat, 100
Technological cycle, 100
Technological forecasting, 18, 19
Technological unemployment, see also Unemployment, 1, 97, 98, 100, 104
Technologically backward social groups, 6
Technologists, 198
Technology
 assessment, 25, 137–39, 192
 assessment in Japan, 124, 125
 a definition of, 26
 distinguished from science, 3
 gap and its disappearance in Japan, 120
 and economic and social ends, 71
 links with science, 76–79
 policy options and priorities, 31, 76
 and pollution, 1
 sequences, 76, 79, 80
 sources of, 77
 transfer, 91–93, 137, 142, 192
 transfer within Canada, 163
 transfer in Japan, 120
 transfer in the Netherlands, 166
Terms of trade and technological change, 41
Teubal, M., 72
Thailand, new technology and economic development, 70
Tisdell, C. A., 26–28, 30, 72–74, 103, 104, 146, 164, 165
TNO (Netherlands Organisation for Applied Scientific Research), 164, 165, 192
Tobin, J., 14, 27, 74
Toffler, A., 99, 105
Toulmin, S., 58, 74
Trade and Industry Organization (Vorort), Switzerland, 180
Trade unions, and technological change, 87
Transfer of technology, see also technology transfer, 76, 77, 93, 96, 100
Transnational companies, see Multinational firms
Treasury Board, Canada, 158, 159, 161
Treub, L., 180, 187
Tullock, G., 25, 30

Unemployment, see also Employment, 162, 189, 198, 199, 201, 203
 and economic growth, 100
 structural, 100
 technological, 1, 97, 98, 100, 104
United Kingdom, see also Britain, 56, 107, 135, 142, 144, 145, 147, 164, 168, 175, 181, 184, 190, 191, 199, 200, 203
 decline in patents filed by individual inventors, 81
 GERD compared to Germany, Japan and USA, 109
 GERD performed by government, 26
 GERD allocated to higher education, 34
 industrial R & D compared to Germany, Japan and USA, 109
 industrialization compared to Germany, Japan and USA, 109
 innovation in industry, 83, 86
 international dependence compared to Germany, Japan and USA, 108
 management, marketing and successful innovation, 86
 R & D effort by socio-economic objective, 132, 133
 regional industrial policies, 98
 size of economy compared to

222 *Science and Technology Policy*

    Germany, Japan and USA, 108
  S & T policies and priorities, 126–34
  structure of employment, 109
  and transfer of technology, 91–93
  and voluntary research associations, 64
United Nations, 69
  Conference on Science and Technology for Development, 134
United States of America, 107, 132, 144, 145, 147, 175, 181, 182, 184, 192, 196, 199, 200
  cost of bringing an invention to market, 93
  decline in patents filed by individual inventors, 81
  economic development, 21
  GERD compared to Germany, Japan and UK, 109
  GERD performed by government, 26
  GERD allocated to higher education, 34
  industrial R & D compared to Germany, Japan and UK, 109
  industrialization compared to Germany, Japan and USA, 109
  innovation in industry, 84
  international dependence compared to Germany, Japan and UK, 108
  nature of R & D compared to Japan, 40
  pressures on Japan, 43, 44
  and product cycle, 43, 44, 199, 200
  size of economy compared to Germany, Japan and UK, 108
  S & T policies and priorities, 15, 134–43
  structure of employment, 109
  technology transfer to Australia, 109
United States Congress, 138–40, 147
Universities, 33, 132, 134, 142, 152, 153, 162, 163, 167, 174, 191, 192, 197
  Belgium, 154, 155
  and community and economic needs, 35
  employment of graduates, 35

  vs government *vs* industry as performers of R & D, 63, 64
  and industry, 131, 138, 180
  inventions and NRDC, UK, 93
  and NRDP, Japan, 125
  teaching and research complementarity, 34
  and technology transfer, 91, 92
University Grants Committee, UK, 127, 129
Urbanization, and technology, 97
USA, *see* United States of America
US Department of Commerce, 93
USSR, size of GERD, *see also* Soviet Union, 110

Vaizey, J., 101, 102
Values, and scientific and technological progress, 34
Veblen, T., 24
Vernon, R., 42, 43, 72, 73
Vintage of machines, and diffusion of technology, 83, 87

Wagner, S., 102
Walker, W., 27, 102
Welfare, *see also* Social welfare, 13, 14
Wells, G.C., 42, 72
West Germany, *see* Germany
Whiting, A., 103, 104
Williams, B. R., 34, 38, 39, 71–73
Work, its distribution, 201
Working conditions
  and science policy, 55
  a S & T priority in Belgium, 155
  a S & T priority in Germany, 114, 115
  and technology, 76, 77, 99
World Bank, 70, 75
Wright, J., 205

X-inefficiency, 88

Young, S., 103

ZWO (Organization for Advancement of Pure Research), the Netherlands, 167